CONWY LLOYD MORGAN

AN INTRODUCTION

TO

COMPARATIVE PSYCHOLOGY

WITH DIAGRAMS

Elibron Classics
www.elibron.com

Elibron Classics series.

ISBN 1-4212-2775-4 (paperback)
ISBN 1-4212-2774-6 (hardcover)

This Elibron Classics Replica Edition is an unabridged facsimile of the edition published in 1894 by Walter Scott, Ltd., London.

The Contemporary Science Series.

I. THE EVOLUTION OF SEX. By Professor PATRICK GEDDES and J. ARTHUR THOMSON. With 90 Illustrations. Second Edition.

"The authors have brought to the task—as indeed their names guarantee—a wealth of knowledge, a lucid and attractive method of treatment, and a rich vein of picturesque language."—*Nature.*

II. ELECTRICITY IN MODERN LIFE. By G. W. DE TUNZELMANN. With 88 Illustrations.

"A clearly-written and connected sketch of what is known about electricity and magnetism, the more prominent modern applications, and the principles on which they are based."—*Saturday Review.*

III. THE ORIGIN OF THE ARYANS. By Dr. ISAAC TAYLOR. Illustrated. Second Edition.

"Canon Taylor is probably the most encyclopædic all-round scholar now living. His new volume on the Origin of the Aryans is a first-rate example of the excellent account to which he can turn his exceptionally wide and varied information. . . . Masterly and exhaustive."—*Pall Mall Gazette.*

IV. PHYSIOGNOMY AND EXPRESSION. By P. MANTEGAZZA. Illustrated.

"Professor Mantegazza is a writer full of life and spirit, and the natural attractiveness of his subject is not destroyed by his scientific handling of it."—*Literary World* (Boston).

V. EVOLUTION AND DISEASE. By J. B. SUTTON, F R.C.S. With 135 Illustrations.

"The book is as interesting as a novel, without sacrifice of accuracy or system, and is calculated to give an appreciation of the fundamentals of pathology to the lay reader, while forming a useful collection of illustrations of disease for medical reference."—*Journal of Mental Science.*

VI. THE VILLAGE COMMUNITY. By G. L. GOMME. Illustrated.

"The fruit of some years of investigation on a subject which has of late attracted much attention, and is of much importance, inasmuch as it lies at the basis of our society."—*Antiquary.*

VII. THE CRIMINAL. By HAVELOCK ELLIS. Illustrated.

"An ably written, an instructive, and a most entertaining book."—*Law Quarterly Review.*

VIII. SANITY AND INSANITY. By Dr. CHARLES MERCIER. Illustrated.

"Taken as a whole, it is the brightest book on the physical side of mental science published in our time."—*Pall Mall Gazette.*

IX. HYPNOTISM. By Dr. ALBERT MOLL. Second Edition.

"Marks a step of some importance in the study of some difficult physiological and psychological problems which have not yet received much attention in the scientific world of England."—*Nature.*

X. MANUAL TRAINING. By Dr. C. M. WOODWARD, Director of the Manual Training School, St. Louis. Illustrated.

"There is no greater authority on the subject than Professor Woodward."—*Manchester Guardian.*

XI. THE SCIENCE OF FAIRY TALES. By E. SIDNEY HARTLAND.

"Mr. Hartland's book will win the sympathy of all earnest students, both by the knowledge it displays, and by a thorough love and appreciation of his subject, which is evident throughout."—*Spectator.*

XII. PRIMITIVE FOLK. By ELIE RECLUS.

"For an introduction to the study of the questions of property, marriage, government, religion,—in a word, to the evolution of society,—this little volume will be found most convenient."—*Scottish Leader.*

XIII. THE EVOLUTION OF MARRIAGE. By Professor LETOURNEAU.

"Among the distinguished French students of sociology, Professor Letourneau has long stood in the first rank. He approaches the great study of man free from bias and shy of generalisations. To collect, scrutinise, and appraise facts is his chief business."—*Science.*

XIV. BACTERIA AND THEIR PRODUCTS. By Dr. G. SIMS WOODHEAD. Illustrated.

"An excellent summary of the present state of knowledge of the subject."—*Lancet.*

XV. EDUCATION AND HEREDITY. By J. M. GUYAU.

"It is at once a treatise on sociology, ethics, and pædagogics. It is doubtful whether among all the ardent evolutionists who have had their say on the moral and the educational question any one has carried forward the new doctrine so boldly to its extreme logical consequence."—Professor SULLY in *Mind.*

XVI. THE MAN OF GENIUS. By Professor LOMBROSO. Illustrated.

"By far the most comprehensive and fascinating collection of facts and generalisations concerning genius which has yet been brought together."—*Journal of Mental Science.*

XVII. THE GRAMMAR OF SCIENCE. By KARL PEARSON, M.A., Gresham Professor of Geometry. Illustrated.

"The problems discussed with great ability and lucidity, and often in a most suggestive manner, by Prof. Pearson, are such as should interest *all* students of natural science."—*Natural Science.*

XVIII. PROPERTY: ITS ORIGIN AND DEVELOPMENT. By Professor LETOURNEAU.

"M. Letourneau has read a great deal, and he seems to us to have selected and interpreted his facts with considerable judgment and learning."—*Westminster Review.*

XIX. VOLCANOES: PAST AND PRESENT. By EDWARD HULL, M.A., LL.D., F.R.S. With 45 Illustrations.

"A very readable account of the phenomena of volcanoes and earthquakes."—*Nature.*

XX. PUBLIC HEALTH. By Dr. J. F. J. SYKES. With numerous Illustrations.

"Takes up essential points in evolution, environment, prophylaxis, and sanitation bearing upon the preservation of public health."—*Lancet.*

XXI. MODERN METEOROLOGY. By FRANK WALDO, Ph.D., etc. With 112 Illustrations.

"The present volume is the best on the subject for general use that we have seen."—*Daily Telegraph.*

XXII. THE GERM-PLASM: A THEORY OF HEREDITY. By Dr. A. WEISMANN. Illustrated.

"There has been no work published since Darwin's own books which has brought to light so many new facts."—*British Medical Journal.*

XXIII. INDUSTRIES OF ANIMALS. By F. HOUSSAY. Illustrated.

"His accuracy is undoubted, yet his facts out-marvel all romance. These facts are here made use of as materials wherewith to form the mighty fabric of evolution."—*Manchester Guardian.*

XXIV. MAN AND WOMAN: A STUDY OF HUMAN SECONDARY SEXUAL CHARACTERS. By HAVELOCK ELLIS. With Illustrations.

"There is no work approaching his [Mr. Ellis's] in exhaustiveness, accuracy, and fairness of judgment."

XXV. THE EVOLUTION OF MODERN CAPITALISM. A STUDY OF MACHINE PRODUCTION. By JOHN A. HOBSON, M.A. With Diagrams.

XXVI. APPARITIONS AND THOUGHT-TRANSFERENCE: OR, THE COMMUNICATION OF SENSATIONS, IDEAS, AND EMOTIONS OTHERWISE THAN BY THE KNOWN SENSES. By FRANK PODMORE. Illustrated.

THE CONTEMPORARY SCIENCE SERIES.

EDITED BY HAVELOCK ELLIS.

COMPARATIVE PSYCHOLOGY.

AN INTRODUCTION

TO

COMPARATIVE PSYCHOLOGY.

BY

C. LLOYD MORGAN,

PRINCIPAL OF UNIVERSITY COLLEGE, BRISTOL.

WITH DIAGRAMS.

LONDON:

WALTER SCOTT, LIMITED,

PATERNOSTER SQUARE.

1894.

CONTENTS.

PAGE

PROLEGOMENA 1

CHAPTER I.

THE WAVE OF CONSCIOUSNESS . . . 11

CHAPTER II.

THE PHYSIOLOGICAL CONDITIONS OF CONSCIOUSNESS 24

CHAPTER III.

OTHER MINDS THAN OURS . . . 36

CHAPTER IV.

SUGGESTION AND ASSOCIATION . . . 60

CHAPTER V.

ASSOCIATION OF IDEAS IN ANIMALS . . 84

CHAPTER VI. PAGE

MEMORY 99

CHAPTER VII.

MEMORY IN ANIMALS 117

CHAPTER VIII.

THE ANALYSIS OF IMPRESSIONS . . . 125

CHAPTER IX.

SYNTHESIS AND CORRELATION . . . 144

CHAPTER X.

THE SENSE-EXPERIENCE OF ANIMALS . . 157

CHAPTER XI.

AUTOMATISM AND CONTROL . . . 173

CHAPTER XII.

INSTINCT AND INTELLIGENCE . . . 197

CHAPTER XIII.

THE PERCEPTION OF RELATIONS . . 217

CHAPTER XIV.

DO ANIMALS PERCEIVE RELATIONS? . . 241

CHAPTER XV.

PAGE

Conceptual Thought 262

CHAPTER XVI.

Do Animals Reason? 287

CHAPTER XVII.

Subject and Object 305

CHAPTER XVIII.

The Evolution of Consciousness . . 323

CHAPTER XIX.

Selective Synthesis in Evolution . . 333

CHAPTER XX.

The Psychology of Man and the higher Animals Compared. . . . 358

Index 379

PREFACE.

MY central object in this work is to discuss the relation of the psychology of man to that of the higher animals, since such a discussion forms in my opinion the best introduction to Comparative Psychology. A secondary object, subordinate indeed, yet forming an integral part of my plan, is to consider the place of consciousness in nature, the relation of psychical evolution to physical and biological evolution, and the light which comparative psychology throws on certain philosophical problems.

It was my original intention to compare my own results with those which have been reached by previous observers and thinkers in this field of investigation and inquiry. But I found that, in the first place, this would largely increase the bulk of the book; that, in the second place, it would introduce a controversial tone, which I was desirous of avoiding; and that it would in other ways interfere with what appeared the most convenient mode of developing my subject. I therefore abandoned my original intention, and adopted a more direct method of exposition and discussion. It is, however, all the more incumbent on me to acknowledge my indebtedness to my predecessors and contemporaries. Those whose acquaintance with the subject is most wide and extensive, will best be able to judge how far what I have written is a mere restatement of what has already been written, and how far, if at all, I have done something towards advancing the boundaries of our knowledge or rendering the knowledge that we possess clearer and more exact. Others will perhaps do well to regard me as a secretary who has, I trust, with due diligence thrown into convenient form the data with which he has been supplied.

It is right that I should particularly call attention to my indebtedness to the works of Mr Herbert Spencer, whose description of relations as "the momentary feelings accompanying transitions" in consciousness contains the germinal idea from which my own treatment of the perception of relations has developed; of Professor William James, whose conception of a wave of consciousness I have adopted; of Professor Romanes,* to whom I owe much, and in many ways; and of Professor Mivart, in whose writings I find many things with which I am in cordial agreement, and not a little from which I must dissent.

In the body of the work there will be found a description of certain experiments and observations on newly-hatched chicks and ducklings, some of which have already been published in the *Fortnightly Review* for August 1893, and in *Natural Science* for March 1894. Similar observations have been made by previous observers, and the results obtained by Douglas Spalding and Professor Eimer are often quoted. In my own work on "Animal Life and Intelligence" I cited some of these results. Shortly

* The death of Professor Romanes, since this too brief acknowledgment of all that I owe to him was written and printed, has entailed a loss to Science which is irreparable, and a loss to his personal friends which lies too deep for words.

after the publication of that book, I received from my friend, Mr T. Mann Jones, a letter informing me that experiments and observations of his own gave different results, and had led him to regard with critical suspicion much that was written about the "philosopher's chick." He especially adduced the instinctive knowledge of the cry of a hawk, as such, said to be shown by young turkeys in the absence of individual experience, as in his opinion an unwarrantable assumption. I therefore determined to carry out a series of observations for myself, especially as I wished to study association in young birds. The results of these observations are, on the whole, confirmatory of those to which Mr Mann Jones's investigation (which I trust he may be induced to publish) had led him. With regard, therefore, to the observations and experiments in this matter described in this book, I would ask the reader to bear in mind, that they are repetitions of similar observations made by many previous observers; that the results to which they lead differ in some respects from those obtained by Spalding and by Professor Eimer; and that they are, in their general tendency, confirmatory of the views which Mr Mann Jones has communicated to me, and of the general conclusions on instinct in young birds advocated by Dr A. R. Wallace, Mr W. H. Hudson, and others.

I have laid much stress on the paramount importance of systematic and sustained observation as the only safe basis for conclusions concerning the intelligence of animals. The observations, hereafter described, of the way in which dogs deal with the difficulty of bringing a stick through vertical railings, affords a case in point. But with regard to this case,

I should here state that I have received communications from two correspondents—Miss M. E. Garnons Williams and "Ouida"—describing observations which show that dogs sometimes meet the difficulty without preliminary bungling. I do not think that such occasional observations invalidate the conclusions reached by systematic investigation. I would, however, urge on all those who have the good fortune to witness the performance of some conspicuously intelligent action in any animal, not to rest content with merely recording it, but to make it the basis of further observation directed to the end of ascertaining its true nature. The records of casual observation are not without their interest; but the results of detailed investigation are, for comparative psychology, of far greater value.

I would strongly urge upon my readers the advisability of testing, by careful introspection, all my statements concerning the mental processes of man. Only thus can a valid basis either of appreciation or of criticism be obtained. It must not be forgotten that introspective psychology is an essential preliminary to comparative psychology, and that, if it is to produce results of scientific value, it must be based upon exact and oft-repeated observation. Such observation, however, requires special training, not less than objective observation in physics or in biology. It would be an inestimable boon to comparative psychology, if all those who venture to discuss the problems with which this science deals would submit to some preparatory discipline in the methods and results of introspective observation.

I have endeavoured throughout to be self-consistent in my use of technical terms. But it will be noticed that some terms, such as "perception" and "conception," "percept" and "concept," are employed in senses somewhat different from those which are commonly accepted; and indeed different from those which I had myself been led to adopt in a previous book. On this matter I would court criticism. In another work, now in the press, entitled "Psychology for Teachers," I have further endeavoured to give clear expression to my views in this matter; and I beg to refer those who are interested in the matter to that work.

I have received friendly help and criticism from many correspondents, to whom I hereby tender my best thanks. Among these I would especially mention Mr T. Mann Jones, some of whose observations on animals are quoted in the Appendix of Mr Herbert Spencer's volume on "Justice." Mr Jones has been good enough to supply me with much information which is the outcome of his wide experience of domestic animals, and also to give me permission to make use of the information he thus placed at my disposal. One of my chief grounds for not availing myself of this privilege, is my hope that Mr Jones may himself be induced to publish a work on animal intelligence, in which these observations would find a fitting place. But I am anxious here to acknowledge how helpful my correspondence with him has been to me in my own work. Another correspondent who has favoured me with helpful criticism is Mr H. B. Medlicott. If the following pages show that I am not able fully to accept the view which he put forward in his pamphlet on "The Evolution of Mind in Man," I can assure him that I am not the less grateful for the elucidation of that view with which he has favoured me in private correspondence. I have also to thank my friends Mr Norman Wyld and Mr Sidney H. Reynolds, my brother Dr Llewellyn Morgan, and Miss Ashe, for the trouble they have taken in reading much of my work in MS. or in proof, and for the valuable assistance they have rendered me.

In conclusion, I desire to express my indebtedness to the Editors and Publishers of the *Fortnightly Review*, the *Monist*, the *New World*, and *Natural Science*, for their courteous permission to introduce parts of articles which have appeared in their pages.

C. LLOYD MORGAN.

BRISTOL, *September* 1894.

COMPARATIVE PSYCHOLOGY.

AN INTRODUCTION

TO

COMPARATIVE PSYCHOLOGY.

PROLEGOMENA.

IN a treatise on human psychology, it may be possible and advisable to proceed on purely empirical lines and to keep in the background the philosophy of the subject. But in a consideration of comparative psychology, which introduces the conception of evolution, and of the relation of mind to the organism, such a procedure seems to be neither possible nor advisable. It will conduce to clearness and prevent misconception, therefore, if I here prefix to my work an introductory section, containing, in the form of *prolegomena*, a succinct account of the monistic views which I accept.* Nearly every philosopher contends now-a-days that he is a monist. But there are monists and monists. I must therefore endeavour to state clearly the form of monism which I accept.

First of all, I accept *a monistic theory of knowledge.* The dualist starts with the conception of a subject introduced

* Those for whom philosophy has no special interest will do well to pass over these *prolegomena* and proceed at once to Chapter I. Those on the other hand for whom the philosophical interest is central should re-read this section after the perusal of the body of the book.

into the midst of a separately and independently existent objective world. For him the problem of knowledge is how these independent existences, subject and object, can be brought into relation. In the monistic theory of knowledge it is maintained that to start with the conception of subject and object as independent existences is false method, and that the assumed independence and separateness is nowise axiomatic. Starting then from the common ground of *naïve* experience, it contends that, prior to philosophizing, there is neither subject nor object, but just a bit of common practical experience. When a child sees a sweet, or when a dog sees a cat, there is a piece of *naïve* and eminently real experience upon which more or less energetic action may follow. It is only when we seek to *explain* the experience that we polarize it in our thought into subject and object. But what logical right have we to say that the subject and object, which we can thus distinguish in thought, are separate in existence? No doubt it is a not uncommon, and a not unnatural, fallacy to endow with independent existence the distinguishable products of our abstract and analytic thought. The distinguishable redness and scent of a rose may thus come to be regarded as not only distinguishable in thought, but also separable in existence. But, until it shall be shown that "distinguishable in thought" and "separate in existence" are interchangeable expressions, or that whatever is distinguishable is also independent, the conclusion is obviously fallacious. And it is this fallacy which the monist regards as the fundamental error of the dualistic theory of knowledge. While dualism, then, starts with what I deem the illegitimate assumption of the independence of subject and object, the monist, starting from the common ground of experience, looks upon subject and object as the distinguishable aspects of that which in experience is one and indivisible. They are distinct from each other, and the distinction is fundamental; but they are nowise independent and separate in existence. The apparent dualism is a dualism of aspect, not a dualism of existence. It need only be added that this is a theory of knowledge, and of the experience of which knowledge is the outcome. Of that which is not known and not experienced, it neither asserts nor denies anything. But, accepting as it does the reality of experience, it does assert that the aspect which we polarize as objective is just as real, and real in the same sense, as the aspect we polarize as subjective. The reality of object and subject is strictly co-ordinate. And those who hold this view regard as little better than nonsense the assertion that whereas the reality of the subject is unquestionable, the reality of the object is a matter that is open to discussion. Self and cosmos are of co-ordinate reality: they are the polarized aspects of experience as explained through reason.

But a theory of knowledge is not a complete interpretation of nature. There lies on a lonely mountain height a stone. I ascend the mountain, and the stone becomes an object of experience. That we may explain on our monistic theory of knowledge. But how about the stone before I got there, and after I had left the mountain top, when it was not yet or no longer an object of experience? To this question the theory of knowledge that is modest and knows its business replies:—"I do not know. I deal with experience. I can tell you nothing concerning that which is not yet experienced or no longer experienced. That is a matter of the interpretation of nature."

There are some excellent folk who believe that a philosophy can be built up without assumptions. I am not among their number. Hypotheses, or assumptions, are as necessary in philosophy as they are in science. I assume that nature is wider than actual experience. I assume that the stone on that lonely mountain top exists in some form capable of again appearing as object, and that this continuous existence is quite independent of whether anyone is there to experience it as object or not. I cannot possibly prove this, and do not attempt to do so. I suppose I accept it for this reason, that of the two hypotheses (*a*) that it continues to exist in some form or other whether it be an object of experience or not, and (*b*) that it dodges in and out of existence according as it is experienced or not-experienced, (*a*) appeals to me as the more rational assumption. Anyhow, if I cannot prove (*a*), neither can anyone else prove (*b*). I assume, then, that the world, which forms the objective aspect of experience, continues somehow to exist quite independently of its being sensed or perceived. That there is a nature to interpret is thus an hypothesis or assumption, the sole justification of which is that, though it can never be proved, it accords more satisfactorily with the facts of experience than any other assumption. It does not conflict with, but supplements the monistic theory of knowledge. It fills in the gaps of actual experience with "permanent possibilities" of experience.

I pass now to a second phase of monism. I accept *a monistic interpretation of nature.* What do I mean by a monistic interpretation? Well, the essence of this view comes out when we consider the position of man in nature. According to this hypothesis, man, as an organism, is one and indivisible (though variously maimable), no matter how many aspects he may present subjectively or objectively. That the inorganic and organic world have reached their present condition through process of evolution, is now widely accepted. But the dualist contends that mind is a separable existence, *sui generis,* forming no part of the natural world into which it is temporarily introduced. Here the monist joins issue, and contends that, alike in its biological and its psychological aspect, the organism is the product of evolution; that mind is not extra-natural nor supra-natural, but one of the aspects of natural existence.

Observe the frankly hypothetical nature of this view. The monist assumes that what we call nature is co-extensive with knowable existence. He assumes that far, very far, as we may be at present from anything like a complete or adequate explanation of nature, yet still this nature is explicable, and that by one method, the method of scientific procedure. Herein lies the essence of monism as an interpretation of nature. If in the wide region of the known and the knowable there be any modes of existence which not only are not explicable, but from their very nature never can be explicable, as parts of one self-consistent whole, our monism falls to the ground. We contend that it is this to which science, philosophy, poetry, ay and religion too, has been tending throughout the centuries of human progress.

A monistic interpretation of nature, so long as it holds true to the main principle of being throughout self-consistent, allows any amount of individual freedom in the treatment of details. It is characterized, not by the possession of a common scientific or philosophic creed, but by a common aim. It appears to me, for example, that in the evolution which sweeps through nature, the process is throughout characterized by the following traits:—(1.) It is selective; (2.) it is synthetic; (3.) it tends from chaos to cosmos. And these traits seem to me characteristic alike of inorganic, organic, and mental evolution. Now I dare say there are not half-a-dozen independent monists who will agree with me in singling out these three traits for especial prominence. But what does that matter? My aim is monistic, as is also theirs. And there is plenty of room for many differences and even divergences of opinion among those who are in search of a self-consistent theory of thought and things.

I now turn to a third aspect of monism, which may be termed *analytic monism.* This consists in an analysis of the object of knowledge, or in other words, of nature as known and knowable. Now here it is essential quite clearly to grasp the fact that all that we know must, in the act of becoming known, be an object of knowledge. The object of knowledge is not merely the object of sense, but includes also the object of thought. All that we know of the subject, all that we attribute to the self, must, in becoming known, be the object of thought. It is only in reflection or introspection, which is also retrospection, that this is possible. You cannot analyse any bit of experience at the moment when it is being experienced, you can only look back upon it in a subsequent moment of reflection. In that subsequent moment it may be polarized into object and subject, and either the objective aspect or the subjective aspect may then be the object of thought. In this way the subjective aspect of experience in moment (*n*) may be the object of thought-experience in any subsequent moment (*q*). But never can the subject of experience in any moment be the object of knowledge in the same moment. Hence it follows that without reflection there can be no knowledge of the subjective aspect of experience. And hence it follows also that our knowledge is always dealing with the self of a moment ago. It is an assumption which can never be proved, but one on the validity of which we all place implicit reliance, that the subject is continuous, and that the subject of the present moment is practically identical with the subject of a moment ago, of which we have knowledge through reflective thought.

Let us now take that natural object which we call a man, and let us assume that he is constituted in all essential respects as we are. We analyse him in thought; and we may carry our analysis but a short distance, or as far as ever we can. Analyse him a little way down, and we reach the conception of body and mind. It is clear that the concepts of this analysis are closely connected in origin with the concepts reached by the analysis of experience, and that body and mind are analogous to the object and subject of sense experience. Now the fact to which analytic monism should, as it seems to me, stick close, is that body and mind are the products of analysis. What is practically given is the man; and this man is one and indivisible, though he may be polarized in analysis into a bodily aspect and a conscious aspect. It may be said that this is an assumption. Granted. It is part of the fundamental assumption of the monistic interpretation of nature. According to that assumption or hypothesis, the organism in all its aspects is a product of natural evolution. We proceed to study that product. We analyse these aspects. We find that a certain group of them hang together in a special way, and we call them bodily aspects; and we find that a quite different group of them hang together in their special way, and we call them mental aspects. There is no getting on without an hypothesis of some kind, and this is the one which the monist adopts. The dualist says that the organism in its bodily aspect is a product of evolution, or of some other process of genesis, and that the mind is implanted therein by some extra-natural process. That is his assumption. The future must decide which assumption is the more reasonable.

According to the monistic assumption, then, the organism is one and indivisible, but is polarizable in analytic thought into a bodily and a mental or conscious aspect. Body and mind, like object and subject, are distinguishable but not separable. And now we may proceed to carry the analysis deeper. We reach the brain, or some part of it; and here our analysis discloses, as one aspect, certain forms of nervous change or transformation of energy, and, as the other aspect, certain phases of consciousness. Note clearly that this is merely through carrying further the same process of analysis; and that, of the products of analysis, neither can claim priority or superior validity to the other. They are strictly co-ordinate: each is as real as the other. The true reality is the man with which the analysis starts: no valid product of the analysis of that man, through the application of rational thought, can be more real than another.

The question then arises: Given an organism in which analysis gives two aspects, complex energy and complex consciousness, from what have these been evolved by an evolution which is selective, synthetic, and cosmic or determinate? From the nature of the case, the bodily aspect is that of which alone we can have objective knowledge. We trace the evolution backwards and find, in our interpretation thereof, simpler and simpler organisms, until the organic passes into the inorganic. We find the energy less and less complex as we look back through the vista of the past. And what about the other aspect? Does it not seem reasonable to suppose that, no matter what stage we select, analysis would still disclose the two aspects? That with the simpler modes of nerve-energy there would go simpler modes of consciousness, and that with infra-neural modes of energy there would be infra-consciousness, or that from which consciousness as we know it has arisen in process of evolution? This is admittedly speculative. But is it illogical?

Let us return, however, from this speculative excursion, to emphasize again the fact that for monism the organism in practical experience is the starting-point; that it is one and indivisible, though it has different aspects, which may be distinguished in analytic thought; and that these aspects are strictly co-ordinate—neither is before nor after the other.

Opposed to such a view are—(1.) the hypothesis of materialism, according to which the body is the real substance, the mind being one of its properties; and (2.) the hypothesis of what may be termed psychism, which is, in the words of Charles Kingsley, that "your soul makes your body, just as a snail makes its shell," that mind is the reality and body the mere phenomenal appearance. These views depart from the cardinal principle of monism, which is that practical experience is the fountainhead of reality. They give to one product of the analysis of this experience a validity superior to that of another product of this analysis.

Now analytic monism by itself is insufficient and partial. It is open to the criticism that, while professedly monistic, it postulates a dual aspect, and is therefore merely dualism in disguise. But this criticism falls to the ground when this analytic monism is taken in association with the monistic theory of knowledge and the monistic interpretation of nature and of man. Monism must be judged as a whole, or not at all. Its cardinal tenets are:—That nature is one and indivisible, and is explicable on one method, the method of knowledge; that experience is one and indivisible, though we may distinguish its subjective and objective aspects; that man is one and indivisible, though our analysis may disclose two strongly contrasted aspects, body and mind. It contends that man in both aspects, biological and psychological, is the product of an evolution that is one and continuous; and, combining the results of its theory of knowledge with those of its analysis of man, it identifies the mind, as a product of evolution, with the subject, as given in experience.

There is one further result of the analysis of experience upon which I must briefly touch in bringing these *prolegomena* to a conclusion. Monism regards nature and experience as one and indivisible, and all apparent dualism, as a dualism of aspect, distinguishable in thought, but indissoluble in existence. It contends, as I have endeavoured to show, that the individual mind on the one hand, and the cosmos on the other hand, are alike products of an evolution which is one and continuous. In both the products which we thus distinguish we find a synthesis which is selective and determinate. Empirically, that is as far as we are justified in going. Empirically we must just accept this continuous and progressive synthesis as the ultimate conclusion of science. But it is characteristic of man as a thinker that he is seldom able to stop here. He is constrained to take one further step in his analysis; and it only remains for me to indicate the nature of this final step as viewed in the light of a monistic philosophy. The selective synthesis of the cosmos, which we call evolution, is regarded as the manifestation, under the conditions of time and space, of an underlying activity which is the ultimate cause thereof. This underlying activity is not a product of evolution; it is that in and through which evolution is rendered possible. In like manner the selective synthesis of my mind, which we term its natural development, is regarded as the manifestation, under the conditions of time and space, of an underlying activity, one in existence with and yet distinct in analysis from that of the cosmos at large. This underlying activity, which is the ultimate essence of my individual personality, is not a product of evolution; it is that in and through which the evolution of my consciousness is rendered possible. Object and subject are thus the correlative modes of manifestation of an underlying activity, one in existence, but none the less fundamentally distinct in aspect.

CHAPTER I.

THE WAVE OF CONSCIOUSNESS.

I DO not propose to begin by defining consciousness. Any definition would be found, when analysed, to involve a direct reference to primary experience. I shall therefore assume that my reader has this primary experience; that he is conscious, and that he knows what I mean when I say that he is conscious. And I ask him to verify in his own experience the preliminary statements which I make in this chapter. The first of these is, that we are only directly aware of *present* states of consciousness. I use the word "present," not in its abstract sense of an ideal boundary between past and future, but as descriptive of a short but not inappreciable period of time. I shall speak of this short period of time as the *moment of consciousness.* This phrase, as used in this work, is descriptive of the short period which constitutes the *now* of consciousness. We are only directly aware, I repeat, of that which transpires within the moment of consciousness. This may, however, seem to be contrary to experience. It may be objected that consciousness deals not only with the present but also with the past and with the future; that this, indeed, is one of its distinctive features. But a little consideration will enable us to see that the past or the future must be made *present* to consciousness through memory or anticipation. The recollection of this morning's breakfast is a present state of consciousness; and the expectation of to-morrow's sunrise is also a present state of consciousness. That is to say, in strict phraseology, consciousness does not deal with the past or the future, but with representations of past or future events. And these representations occur in the present moment of consciousness.

The next thing to notice is, that in each present moment there is sequence and change. The question has been raised whether, in the absence of such change, we can be conscious at all,—whether such change is not one of the primary conditions of consciousness. We need not here discuss this question. It is sufficient that for ordinary practical experience, in any moment of consciousness psychical states are constantly coming into being, and passing away, and are continually changing as they pass. Repeat aloud some familiar lines of poetry, and stop suddenly after any word in the course of your recitation. You will be conscious of the last words you have uttered just fading away, and of new words, not yet uttered, but, as we say, on the tip of the tongue, just coming into consciousness. So too in hearing a familiar piece of music, one feels the coming harmonies before they are played. Fix your eyes upon any word near the middle of this page; that word is sharp, clear-cut, and well-defined in vision. Other words near it, above and below as well as on either side, are visible, but not clearly defined. The rest of the page is also visible, and probably a good deal more besides, but only dimly and in hazy outlines. Now read on *slowly*. Your consciousness will pass on from word to word; but you will, I think, be aware of the coming words, which are, so to speak, dawning upon consciousness, and of the words you have just read waning out of consciousness. Ask some one to read aloud *rapidly* from a well-printed page, and as he does so slip a piece of blank paper over the text. He will read six or seven words beyond that which he was pronouncing when the page was covered. In turning over for a pianist, you must do so more or less in advance of the chords he is actually playing.

The moment of consciousness thus embraces a psychical

wave with *a summit or crest of clear consciousness*, a short rising slope of dawning consciousness, and a longer falling slope of waning consciousness. This is very diagrammati-

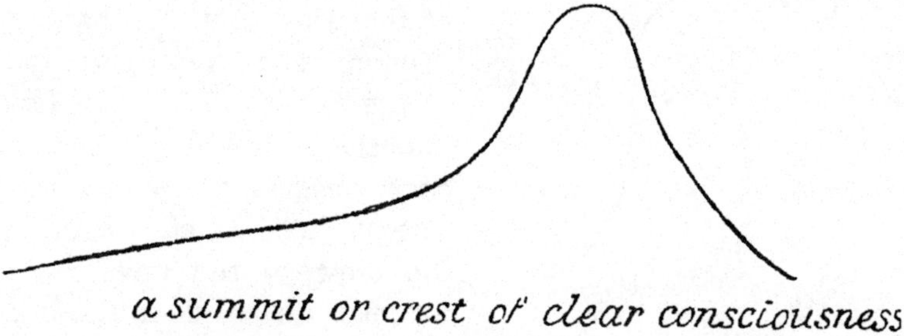

FIG. 1.

cally represented in Fig. 1. The wave of consciousness is supposed to be proceeding from left to right along words printed below it; and when caught by the instantaneous photography of thought, the word "clear" occupies the

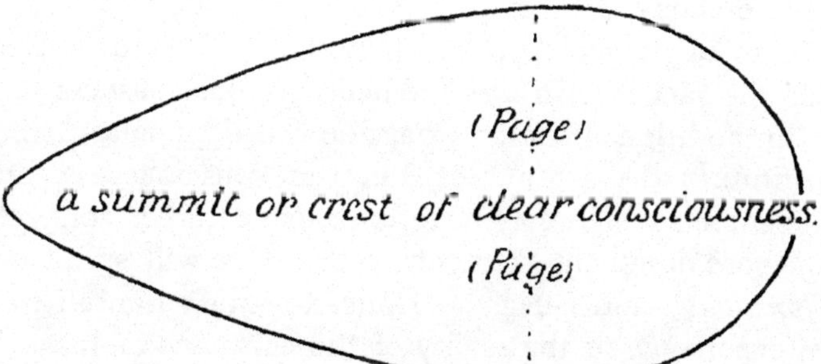

FIG. 2.

summit of the wave; the word "consciousness" is just rising or dawning; and the words "a summit or crest of" are waning out of consciousness. This is a longitudinal section of the wave. Fig. 2 represents a plan of the wave; and if we take a transverse section at right angles to the line of flow (along the dotted line in the plan), we shall have that which is represented in Fig. 3. Here we do not of course see the waxing and waning of the constituent parts of the wave, but we have a diagrammatic representation of the fact that there are less intense lateral elements in consciousness, that exist alongside of those which constitute the crest of the wave. As we read, the words in the line our eye is following occupy successively the summit of the psychical wave. In Fig. 3 the word "clear" occupies, as before, the summit of the wave. But our eye is also being affected by the lines above and below, and by the rest of the page. We do not generally notice this, because it is of no practical importance in everyday life. For us, as students of psychology, it is important. We must clearly grasp the fact that in any moment of consciousness, there are in addition to and alongside the dominant elements constituting the summit of full clear consciousness, dimly felt elements which may have little or no direct connection with those dominant elements. These we will speak of as *subconscious;* and I beg the reader to satisfy himself in his own experience of the reality of the existence of these subconscious elements. As I write, I am dimly aware of the ticking of the clock, of the plash of rain against the window, of the position of my body, of the pressure of my clothes, of the scent and taste of a cigarette, of an incipient headache, and of much besides. All these are merely subconscious, and subconscious in different degrees.

FIG. 3.

As the validity of many of the arguments in this book depend upon the real existence of subconscious elements in the psychical wave, I will further illustrate my meaning in another way and in other phraseology, which, together with that of "the wave of consciousness," I shall have frequent occasion to employ. My further illustration is that from vision,—in most of us the dominant sense. If we fix our eyes on any distant object, such as a church spire or clump of trees, this is in the focus of vision; but it is set in the midst of a wide visual field. The focus shades off into and is surrounded by a margin, in which the objects, instead of being clear-cut and well-defined, like the church spire or the clump of trees, are dim and blurred in outline. The focus here answers to the summit or crest of the psychical wave; the margin to its subconscious body, comprising all the rest of the wave other than the crest. Now, although this illustration is based on vision, it is applicable to consciousness generally. Those who have an ear for, and some little knowledge of, music, can, when they are listening to a four-part song, focus their attention on the treble, alto, tenor, or bass, making that the dominant theme, and allowing the other parts to be marginal. For the ordinary listener, the air is focal, the other parts marginal. Music, which so often gives us a leading theme, vocal or other, and its setting or accompaniment, affords indeed an excellent illustration of what I am seeking to enforce,—that in addition to what is focal in consciousness (the theme) there is much that is subconscious or marginal (the accompaniment). And this I repeat is true not only of vision or of hearing, but of consciousness in general. I take a walk in the country with a friend, and we are discussing the relation of music to poetry. This is our theme; it is focal to our consciousness; the points in the discussion as they arise successively occupy the crest of the psychical wave. But there is plenty of accompaniment; there is much body to the wave. A thousand sights, scents, sounds, all breathing the life of spring; the soft yet invigorating air; the tingling of the muscles to the exercise,—all these form a delightful setting to the theme, a margin to the focus, a body to the wave of consciousness. And any one of these subconscious elements may by increased intensity rise at any moment into dominance, and occupy the crest of the wave or the focus of consciousness. My friend, for example, who is a keen naturalist, stops in the middle of his sentence to point out some rare bird which has caught his attention. Often when one is reading or listening to a discourse, there is an undercurrent of subconsciousness, wholly unconnected with the subject of the book or the lecture; presently this may rise unbidden, and monopolise the wave crest; and one finds that the words over which the eye has been travelling, or which have been falling upon the ear, have suggested nothing definite and rememberable. They have merely entered the margin of consciousness, the focus being otherwise occupied.

In riding a bicycle slight movements of the handle are constantly necessary. But the skilful rider spinning along a good road guides the machine automatically or unconsciously, as we are wont to say,—subconsciously, as we should more correctly say. In all matters of skill, when it is well-established, the mere carrying out of the skilled action is marginal and subconscious, focal consciousness being concentrated on the end to be attained, or on some particular factor in the process. The swordsman who had to focus his attention on, and exercise fully conscious control over each several parry, would soon succumb to his better trained antagonist in whom all this is a matter of organised habit carried out subconsciously. His attention is focussed on his adversary's sword-point. I would ask the reader, who is a billiard-player, or cricketer, or a player of lawn-tennis—in fact of any game requiring skill—to exercise a little self-observation in the matter. He will, if I mistake not, be able to verify the truth of my assertion, that his focal consciousness is constantly fixed on some salient point of the skilled action, while the application of the skill to that salient and focal point is marginal; and, further, that this marginal exercise of the skill is not really unconscious, but is subconscious. He who kicks a goal at football rivets his attention on striking the ball precisely there and thus; all the rest is marginal to his consciousness.

Note then the complexity of the wave of consciousness. We are too apt in psychology to pay attention solely to focal consciousness, omitting all reference to the great body of marginal subconsciousness. But this is a great mistake. The focal consciousness very often is what it is in virtue of the subconscious margin in which it is set. The dawning elements of the psychical wave, the waning elements, and all the marginal elements, form parts of any present state of consciousness, and are more or less instrumental in determining its nature. I shall employ the phrase "state of consciousness" to describe all that is comprised in the psychical wave in any moment of consciousness. This embraces not only the focal constituent, but a greater or less number of marginal constituents, which form the peculiar setting of the focus in the moment of consciousness in question. At different times in the same individual, and presumably in different individuals, the states of consciousness whose succession constitutes the onward-flowing wave, vary considerably in complexity and in intensity. In moments of quiet concentrated thought the states of consciousness are relatively homogeneous and simple. In moments of distraction and of bustling excitement they are heterogeneous and complex. When we languidly let our thoughts wander hither and thither without aim or purpose, the states of consciousness are of low intensity; but in moments of keen excitement, of breathless interest, or of strenuous thought, the intensity is much increased. In the psychical wave, intensity may be represented by the height of the wave-crest above the base line which is termed the "threshold of consciousness;" complexity, by its width, or the number of constituent elements in the state of consciousness embraced by the wave.

The phraseology in which I have endeavoured to express some of the observable facts of consciousness, is based on that method of diagrammatic or graphic representation which has been found of conspicuous assistance wherever it has been employed; I mean the use of curves. And now we may proceed to give to our curve its simplest geometrical expression. In Fig. 4, the horizontal line represents the threshold of consciousness, the vertical line a scale by which the intensity of focal or marginal elements in

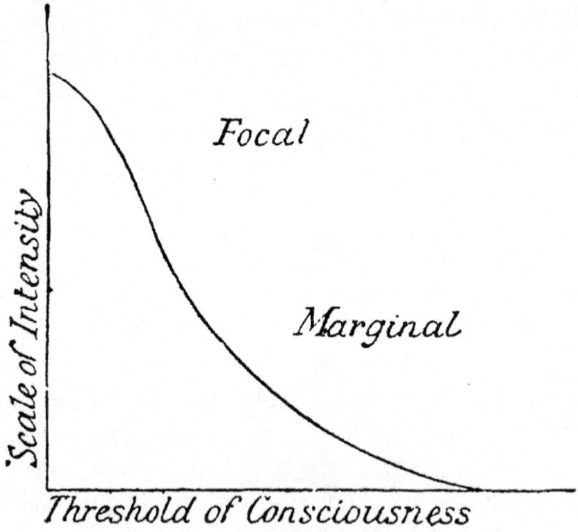

FIG. 4.

consciousness may be measured. The several elements which go to the composition of a state of consciousness are thus spread out along the curve in order of their intensity, from the focal constituents of maximum intensity to the marginal elements near the extreme edge or point of origin of the curve, hardly rising at all above the threshold of consciousness. In this way it is ideally possible to plot down the curve of consciousness. And if in the present condition of our knowledge, and for reasons which will be given directly, it is practically impossible to represent with anything approaching accuracy either the form of the curve or its conscious content,—so complex and difficult of analysis is a state of consciousness,—yet this does not show that the method of representation is false in principle. The question is not whether the method can at present be applied with complete exactitude, but whether as a method it is right in principle. I am convinced that in principle it is right. I believe that, if we fail to recognise that there is such a curve of consciousness, that there is a margin to consciousness as well as the focus, we shall find that the solution of some of the problems of psychology presents difficulties which are almost if not quite insuperable.

But why cannot we apply the method with anything approaching accuracy? In the first place, it is difficult to estimate focal intensity; in the second place, it is quite impossible, I think, to estimate the relative intensities of the marginal elements. Directly we endeavour reflectively to bring any portion of the hazy marginal region into the clear light of focal consciousness, we inevitably find that the measure of our success is also the measure of our failure. Directly we begin to examine and measure any part of the margin, it thereby ceases to be marginal and becomes focal. In the full light of clear consciousness, it must needs seem other than it was when it lurked in the dim shadowy region of subconsciousness. We shall find this a great source of difficulty in some of the investigations into the nature of our conscious states, which are to follow in this book. Moreover, the state of consciousness as experienced is a complex synthetic product not yet analysed. The psychical wave in the moment of consciousness is one and indivisible. The focal element, the dawning elements, the waning elements, the subconscious accompaniment, all fuse into *one state of consciousness*, from which no element could be omitted without altering its identity and making it other than it is. It is only by a process of *introspection*, or looking within at the workings of our own consciousness, that we can gain any direct knowledge of psychical processes. But all such introspection is also *retrospection*. We cannot examine the psychical wave as it passes; we can only endeavour to focus it, or its constituent parts, in the mental vision, *as it was when it was passing*. And here not only is memory apt to play us tricks, but, as before noticed, the act of focussing a marginal constituent thereby makes it other than it was.

The constant changes which the psychical wave undergoes increase the difficulty of accurately representing it at any one phase of its unceasing onward progress in time. For the wave never pauses, but must ever pass on through new changes to new developments. As we read the page of a book, fresh words and ideas are successively dawning, rising to the summit, and waning. The same is true when we throw aside the book, and abandon ourselves to a train of reverie. The wave of consciousness constantly flows on, lifting now these, now those, representations of past or future events to its summit, and letting them sink gently down its backward slope into oblivion. Even when we look moodily out of window, and watch the rain falling continuously against the same dull background of leaden cloud, the wave of consciousness is not arrested. We merely place ourselves under such conditions that the wave of consciousness during succeeding moments remains practically identical in content. The state of consciousness remains the same, only in the sense that the rain seems to remain the same because there is a constant succession of similar raindrops.

There is one more characteristic of the wave of consciousness that remains to be noticed. We have seen that the wave is complex, with focal and duly subordinated marginal constituents; and we have seen that in its ceaseless onward progress in time it is continually undergoing change, so that it seldom remains identical in content for more than a few moments in succession. We have now to draw attention to the fact that in all its changes during its onward course in time it preserves its unity or individuality. Continuity of the psychical wave is unquestionably a characteristic of our conscious experience, which must neither be neglected nor slurred over. In what then does this continuity consist? When we are reading rapidly and with interest, or when in reverie we are reviewing an exciting or amusing scene, the focal constituents of the wave are constantly changing. It is not here, I think, that we must look for that which gives continuity to the wave as a whole. It is rather in the marginal body of the wave that we should seek those relatively abiding elements which are carried on from one moment into the next, and so on through a whole series, and which thus serve to link the successive phases into a continuum. For what is necessary to give continuity to anything which is undergoing continual transformation is, that amid the successive changes of certain parts other parts are relatively constant and abiding. Now there would seem to be two sets of elements which contribute to the relatively permanent, and abiding subconscious body of the wave of consciousness. First, there is a group of subconscious elements arising out of the organic condition of the tissues of the body, which, although it varies according as we are healthy or unwell, fresh or tired, bright or depressed, yet retains a considerable amount of uniformity. While the focal consciousness of the crest of the wave is constantly changing, this organic contribution to its marginal body retains a constancy sufficient to form a partial basis of continuity. Secondly, there is a group of subconscious elements arising out of our intellectual and moral existence. Among these are our settled purpose in life, the ideal to which we would attain, our fixed beliefs, and healthy fundamental prejudices. All these *in man* contribute very largely to that continuity of conscious existence of which every one of us has daily experience.

There is, however, another important mode of linkage of state to state in consciousness. We have seen that a more or less rapid succession of focal elements occupy successively the crest of the wave; and it is clear that when any one of these is succeeded by that which follows, it ceases to exist *as a focal state. Bnt it does not disappear out of consciousness.* It is carried on as a marginal element. This we may represent diagrammatically thus:—

A B C D E F, &c.
a b c d e, &c.
a b c d, &c.

Here A B C are successive focal elements. But A when it ceases to be focal is carried on as a, and *a*, into the marginal region of consciousness. So that when E is focal, d and *c* are still hovering in the margin, and tending to give continuity to the phases in which C D and E are focal. D E F, E F G, and so on, are similarly bound together; and in general, throughout any series of successive phases, continuity is maintained by the constant carrying on of focal elements into the margin. This diagrammatic representation is, of course, far simpler than even our simplest experiences. When we are reading a paragraph, we carry with us to the end the gist of the preceding sentences. In the climax of an interesting play, we retain in the margin of our consciousness the main drift of all the preceding scenes. States of consciousness are enormously complex, and there is room for much marginal matter in the broad body of the psychical wave.

Enough has now been said to give a preliminary conception of what has been termed the wave of consciousness. I am desirous that it should be clearly grasped how very complex a state of consciousness is; and, yet with an orderly complexity, such that we might express the relative intensities of the constituent elements in a curve. These complex states are constantly changing as the wave of consciousness flows onwards in time; but amid all these changes there is a continuity, the nature of which I have endeavoured to explain. The explanation as it stands may very likely seem inadequate. But we have at present only just made a beginning of our study of the wave of consciousness.

CHAPTER II.

THE PHYSIOLOGICAL CONDITIONS OF CONSCIOUSNESS.

IT is now a matter of familiar knowledge that the living body contains, and is largely composed of, a complex substance or group of substances called protoplasm, which is found in all the tissues, such as muscle, bone, or nerve. The protoplasm is found in minute particles of various shapes, termed cells; and of these cells or their products all parts of the body are formed. Within the cells, during life, chemical changes of a special and complicated nature take place; and associated with these chemical changes there are transformations of energy, of great delicacy and complexity. But notwithstanding the special nature of the protoplasmic substance, and the complicated character of the ceaseless transformations of energy, it nevertheless remains true that, so far as the body is concerned, all that it does or suffers belongs to the category of occurrences in the material world,—the world with which physical and physiological science have to deal. From the physical point of view life is an orderly, or perhaps we had better say co-ordinated, sequence of transformations of energy. In the living body a particular kind of matter is thrilling with a peculiar and complex series of molecular changes. The special conditions under which a living organism continues to perform its proper activities are sometimes spoken of as the conditions of vitality.

When the body dies nothing material is taken from it, but the orderly sequence of transformations of energy ceases. The co-ordinated chemical and physical changes which are characteristic of life stop; the movements are no longer seen, the conditions of vitality no longer obtain. For a while the substance seems to undergo no obvious change, but then decay sets in; the elaborate chemical materials undergo decomposition, and the body moulders away. The products of this decomposition are still material; and the mouldering of the body neither adds anything to, nor takes anything from, the world's store of matter. And though the orderly sequence of transformations of energy ceases at death, and gives place ere long to the new series of changes which accompany decay, nothing is abstracted from the world's store of energy, nothing annihilated. The death of the body is a change of state or condition of its substance, and its decay is a further change of state or condition of its constituent molecules. Thus the living body, so far as its matter and its energy is concerned, belongs to the physical world.

Accompanying some of the transformations of energy, or molecular changes in the body or some part of it, there are states of consciousness. These must be carefully distinguished from occurrences in the physical series of events. Imagination and fancy, love and hate, shame and remorse, admiration and pity, judgment and inference,—these are psychical states, not physical events. Each of us experiences them for himself. However convinced I may be that my neighbour experiences states of consciousness similar to those of which I am myself aware, I have no direct and immediate acquaintance with any consciousness but my own. It is with such states of consciousness that psychology has to deal. But inasmuch as these psychical states are associated with a physical organism, comparative psychology must pay attention to the nature and conditions of the association.

Grouping together sequent states of consciousness under a comprehensive term antithetical to "the body," we frequently speak of them as pertaining to "the mind." What is this mind? The answer of empirical psychology to this question is: The wave of consciousness constitutes the mind. From other standpoints this answer may be imperfect or incomplete, but for descriptive or empirical psychology it is sufficient. All that we have direct experience of is the psychical wave in the moment of consciousness. When once it has passed, any particular phase of the psychical wave has no more existence than last year's roses, or than the air waves started by my voice five minutes since. Its effects may remain, but itself has ceased to exist.

Of course, in reflection on my own past life, and on the lives of those around me, I apply the term "mind" not merely to the psychical wave at any moment of consciousness, but to the wave of consciousness in its totality. I thus speak of the products of Shakespeare's mind, the development of my children's minds, the depravity of Nero's mind, and so forth. There is nothing in this usage, however, at variance with the view that the wave of consciousness constitutes the mind, and that when the moment of consciousness has passed, the particular phase of the psychical wave which constituted the mind at that moment has no longer any existence, though its effects may remain.

We sometimes speak of ideas being retained in or by the mind. But it is clear that on the psychological view here put forward, this is not an accurate mode of expression, though it suffices for the needs of practical intercourse. Ideas cannot be retained in or by phases of the wave that have ceased to exist. Nor assuredly are there retained in any existing phase all the ideas of all pre-existing phases. Were this so, the wave would soon reach a state of complexity, which would not only be hopelessly bewildering in the moment of consciousness, but would assuredly defy any attempt at psychological analysis. Instead of this, we have to "recall to mind" the ideas which characterised pre-existing phases. This phrase "recall to mind," seems to imply that the ideas which are thus brought to mind in remembrance have to be fetched from somewhere outside the mind. But ideas are states of consciousness. How then can they exist outside the mind? But *do* they so exist outside the mind? The negative answer which we must give to this question seems at first sight to increase our difficulties, but in reality paves the way for their removal.

I said just now that when once it has passed, any phase of the psychical wave no more continues to exist, as such a phase, than last year's roses, or than the tones of my voice which made the air vibrate five minutes ago. But if the rose-tree which bore the roses shall continue to live and to flourish, fresh roses may bloom next summer; and if I and my vocal chords remain, the tones of my voice may again disturb the air. We do not say that last year's roses are retained in the rose-bush to bloom again next year; but we say that so long as the tree lives and is healthy, roses will be reproduced under appropriate conditions. So too in the case of reproduction of ideas. So long as I live and am healthy, ideas may be reproduced under appropriate conditions. In a word, we seem to be thrown back upon organic conditions as the most probable basis for an explanation of the so-called retention of ideas. The ideas as such have ceased to exist; but the organic structure has been modified in such a way that under the needful conditions similar ideas will be again produced. The reproduction of ideas is thus, on this hypothesis, associated with an organic process; and hence we may say that one of the conditions of the orderly manifestations of consciousness is the maintenance of the integrity of the bodily organisation.

There can be little question that in man and mammal the brain, or some part of it, is the specialized seat of consciousness. From all parts of the surface of the body, from eye, ear, nose, and palate, from the muscles, joints, and internal viscera, there run nerves—ingoing or afferent nerves—which eventually are in communication with the brain. These nerves are the channels along which waves of molecular change (impulses) may be transmitted to the nerve-centres. At their outer ends there are special cells which are so delicately constituted that their molecular equilibrium is readily upset. In the retina of the eye there are cells which thrill to the impact of waves of light; in the ear are cells which respond to the sound-vibrations of the air; in the nose and mouth are cells which have their equilibrium disturbed by the vibrations which we interpret as smells and tastes; in the skin are cells which respond to the contact of solid or fluid bodies, and others which thrill to the stimulation of heat or cold. When any of these are stimulated, impulses are propagated along the nerves with which these cells are connected, and thus disturbances are set up in the nerve-centres of the brain. As the result of these disturbances, other impulses are transmitted down other nerves—outgoing or efferent nerves—to muscles which are thus stimulated to contraction, or to glands which are stimulated to secretion.

It is generally believed, as the result of many observations and experiments, that no consciousness accompanies the transmission of impulses along the nerves; but that when the molecular thrill reaches the brain, or some part of it (probably the cerebral hemispheres), *there* consciousness emerges. One of the conditions of consciousness, therefore, seems to be the occurrence of certain molecular vibrations or disturbances in certain specialised nerve-centres within the cerebral hemispheres. The fact that it takes an appreciable and measureable time for impulses to be transmitted, is prettily illustrated by a series of instantaneous photographs, in Mr Muybridge's collection, of a girl pouring ice-cold water over another girl sitting in a bath. In the first photograph, the girl stands with the bucket, and is just beginning to pour. In the second photograph she has emptied the bucket. The girl in the bath is quite surrounded and covered with ice-cold water. But she has not moved a muscle. She is sitting in exactly the same attitude as in the first photograph. The ice-cold water has stimulated the nerve-ends; but no contraction impulse has yet reached the muscles. In a third photograph the girl is leaping from the bath. Her face appears to be the index of somewhat forcible states of consciousness, the *iced* water being somewhat unexpected.

The rate of transmission of the nervous impulse has been the subject of much careful measurement. It travels along a nerve at the rate of about 100 feet in a second. But if it has to pass through the brain the rate is slower; and if there is choice involved, much slower still.

What may be the nature of the connection between the molecular changes in these nerve-centres and the states of consciousness which accompany them is a problem which has been eagerly discussed. According to the view which is most commonly held, and which is taught to most of us in childhood, before we are at all capable of understanding the nature of the problem, the mind and the body are quite different and eventually separable existences. The body is the mere machine in and by means of which the mind works. The mind, therefore, animates the body, and plays the part of engineer to the organic engine. But during life the nature of the connection is such that the mind, though it uses and must use the body as its instrument, is constantly hampered by its association with gross matter; and death at last sets free the mind or soul from the restrictions of the flesh. Now it is not my province here to deal with this matter from any other point of view than that of empirical psychology. From this point of view the molecular changes in the nerve-centres are merely the concomitants or physiological conditions of the mental processes. Thus in common parlance the mind is said to act upon the body through the nerve-centres, and the body to react on the mind through the same instrumentality. And if it be held that, during life, *all* mental processes have their physiological concomitants, it is clear that these physiological concomitants, namely the molecular changes in nerve-centres, would, if completely ascertained, afford an accurate index of the mental processes. For psychology, then, this dualistic solution of the problem comes to this, that physiological and psychical processes are distinct and separate modes of existence, but that during life every change of the one is the strictly parallel concomitant of a change of the other.

According to another, the monistic, solution of the problem, the physiological and psychical processes, though distinguishable, are indivisible. They are not separate existences temporarily associated during life, but different ways of regarding the same natural occurrences. Whereas according to empirical dualism, the curve by means of which we might represent any given state of consciousness is precisely similar to a second curve by which we might represent those concomitant brain-changes, whatever they may be,* which are assumed to be the invariable accompaniments of the state of consciousness; according to monism there is but one curve, which, regarded from one aspect, is psychical, and regarded from the other aspect is physiological. It is clear that, as a matter of psychological interpretation, the differences between these two solutions, which are symbolized diagrammatically in Fig. 5, are not worth quarrelling about, however important they may be as a matter of general philosophy.

Adopting this hypothesis, then, the curve which represents a state of consciousness may also be taken to represent

* As this work does not deal specially with physiological psychology, the little knowledge that we have on this matter cannot here be discussed, and the similarity (or identity) of the physiological and the psychical curve must be left as an assumption.

a co-existent state of physiological change which exists coincidently in the brain. Let us however be quite clear as to what is meant by such a curve.

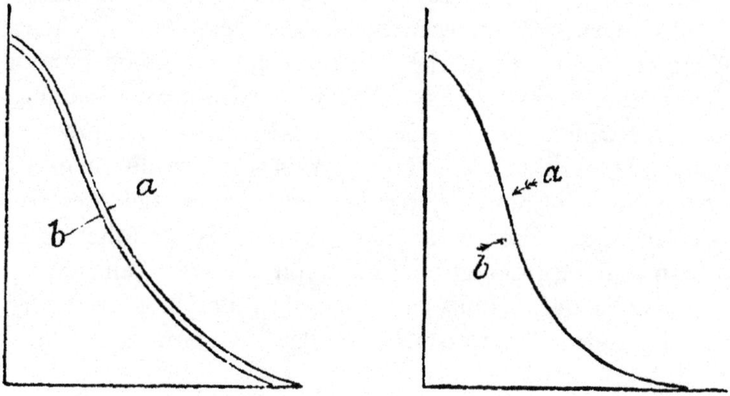

FIG. 5.

A. DUALISM. *a. Psychical Curve.* *b. Physiological Curve.*
B. MONISM. *a. Psychical aspect of the Curve.*
b. Physiological aspect of the Curve.

At any moment of our waking life there are pouring in upon the nerve-centres of the brain afferent impulses, transmitted from all the sensitive parts of the body through the ingoing nerve-channels. When these impulses reach the lower nerve-centres they are co-ordinated, and their net results are, or may be, handed on to the cerebral hemispheres, where they give rise to molecular disturbances, some of which are conscious or are accompanied by consciousness. The nature of these molecular disturbances, as such, it is not necessary that we should here discuss. They may be complex vibrations, unaccompanied by chemical change; more probably they involve changes of chemical state with associated transformations of energy. I shall therefore speak of these as molecular disturbances, or as transformations of energy, without attempting further to particularize their nature. One characteristic of physiological response to stimulus may however be noticed, since it serves to illustrate the physiological aspect of the wave of consciousness. It has recently been very beautifully demonstrated in the electrical discharge of the torpedo-fish, but holds good in many other cases.* The result of a particular stimulus is an effect which is not instantaneous, but lasts an appreciable and measureable time. During this appreciable time it is not uniform in intensity, but at first rapidly increases to a maximum, and then more slowly diminishes. In other words, a graphical representation of the physiological effects of a stimulus is quite similar in general principle to the curve that I gave in Fig. 1, as a longitudinal section of the wave of consciousness.

Now the mass of nerve-centres in the brain is very considerable, even if we exclude all but the highest centres of the cerebral hemispheres; and during waking life the amount of molecular disturbance arising from the impulses coming in through a great number of afferent nerves, which are stimulated in different degrees, must be very great. Moreover, owing to the different degrees of stimulation, and the different sensitiveness of the various parts stimulated, the molecular disturbances in the cerebral hemispheres must vary in intensity. The higher brain-centres may be likened to a piano, in which a number of stretched wires may be set in vibration with different intensities. The keys are the sensitive ends of the afferent nerves, on which the objects round us play, now in one way and now in another. The character of the musical chords evoked will depend in part upon which strings are set in vibration, and in part upon the relative intensities of the vibrations so caused; so

* In vision there is a curious and interesting reduplicated effect or double wave, with a sinus or trough between the two crests, which has recently been investigated by M. Charpentier.

too the character of the states of consciousness evoked through molecular disturbance, will depend in part upon what areas of the nerve-centres are disturbed, and in part on the relative intensities of the disturbances due to afferent impulses.

If then we could represent graphically these relative intensities of molecular disturbance in the cerebral hemispheres, we should obtain a curve of molecular disturbance, which, on the hypothesis of scientific monism, would be identical with the curve of consciousness; or, on the hypothesis of empirical dualism, would accurately correspond to that curve. The dominant disturbances would be those which are fully conscious or focal, and at a lower level on the curve would be the subdominant disturbances which are subconscious or marginal.

But it is exceedingly probable that there are disturbances in the cerebral hemispheres which, though they are of the same order as those which are conscious or subconscious, are of intensity too low to enable them to enter consciousness at all. These we may term infra-dominant. They lie in the region below the threshold of consciousness. On the hypothesis of empirical dualism, the curve of consciousness, which runs parallel with the curve of molecular disturbance down to the threshold of consciousness, there abruptly ceases. But what shall we say of this infra-dominant part of the curve on the hypothesis of scientific monism, according to which the curve of molecular disturbance and that of consciousness are identical, are in fact the same curve regarded from different aspects? If we say that in this region below the threshold there is a form of consciousness which is not even marginally subconscious, we seem guilty of a contradiction in terms; for then we must speak of unconscious consciousness. I shall endeavour to avoid the difficulty, and to leave the matter open for future consideration, by restricting the word consciousness to that which lies above the threshold, and by speaking of that which lies below the threshold as infra-conscious. This infra-consciousness is, in my view, not merely negative, but something positive and existent,—what, for want of better terms, we may call the not-yet or the not-quite conscious; too low in intensity or in kind to become even marginally subconscious, and yet of the same order of existence as that which lies above the threshold. But although this is my own view with regard to the infra-conscious part of the curve, the word I employ does not necessarily imply such positive existence, and may be read by those who so prefer as simply below the level at which consciousness comes into existence.

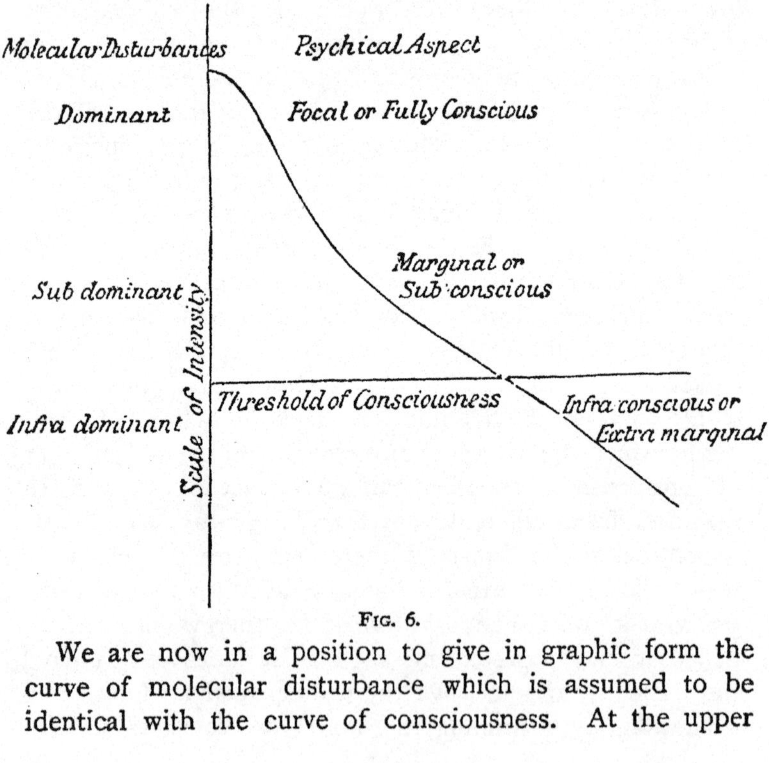

FIG. 6.

We are now in a position to give in graphic form the curve of molecular disturbance which is assumed to be identical with the curve of consciousness. At the upper part of the curve we have the dominant molecular disturbances, which are psychically fully conscious or focal; beneath this are the sub-dominant disturbances, which are psychically subconscious or marginal; and beneath this again are the infra-dominant disturbances which lie below the threshold of consciousness in the infra-conscious or extra-marginal region.

Clear, full, or focal consciousness is thus, on the view here taken, the concomitant of dominant cerebral change. And now the further question arises:—What are the determining conditions of dominance? Obviously this question must be answered separately from the physiological and the psychological aspects. The conditions of dominance of cerebral changes, as such, is a matter of nerve-physiology, and cannot here be discussed. The conditions of dominance of focal states of consciousness—impressions or ideas—is a matter of psychology, the discussion of which we shall take up in the fourth chapter.

It only remains here to add a few words to what was said in the last chapter on the continuity of the wave of consciousness, which constitutes what for empirical psychology is the mind. From the biological point of view, the continuity of mind is one aspect of the continuity and integrity of the living organism. It is clear that the organic contribution to the marginal body of the wave is dependent on the integrity of the organism; and if the intellectual and moral contribution is due to elements of consciousness associated with cerebral changes in the grey cortex of the cerebral hemispheres, it is clear that these also are dependent on the integrity of the organism of which the brain and its cerebral cortex is a constituent part. If this be so, there is no discrepancy between the biological and the psychological conception of continuity; and combining the two, we may say that the continuity which is characteristic of our conscious experience, is conditional on our continued existence as living organisms.

CHAPTER III.

OTHER MINDS THAN OURS.

A DISTINGUISHING feature of modern psychology is the employment of the comparative method. So long as the psychologist restricts himself to the introspective study of the workings of his own consciousness, his conclusions rest on a basis which, however sure it may appear to himself, must be limited by the inevitable restrictions of his own individuality. When he compares and correlates his own results with those of other introspective observers, he becomes so far a comparative psychologist, and by widening his basis renders his conclusions more comprehensive. A further stage of the comparative method is reached, when he endeavours to correlate the results of introspective psychology with the conclusions reached by the physiological study of those nervous processes which are the concomitants of psychical states. On the hypothesis of monism, he is thus comparing two wholly different aspects of the same natural occurrences; on the hypothesis of dualism, two wholly different occurrences, which are nevertheless invariably associated. In any case, by proceeding to this comparison, he links his subject with the science of biology in a way that has proved eminently helpful to his own branch of study. Now, the keynote of modern biology is evolution; and on the hypothesis of scientific monism here adopted, though not necessarily on that of empirical dualism, we are not only logically justified in extending our comparative psychology so as to include within its scope the field of zoological psychology, but we are logically bound to regard psychological evolution as strictly co-ordinate with biological evolution.

I propose to consider in this chapter what we can know of other minds than ours, and how we may gain this knowledge. It follows from what has just been said, that since biological evolution has given rise to individuals of divergent types of organic structure, there may be—nay, there must be—in these divergent biological individuals divergent types of mind, using the word "mind" in the widest and most comprehensive sense as embracing all modes of psychical activity. The question arises, however, how we are to gain acquaintance with these divergent types of mind. And here we are met by the fundamental difficulty which comparative psychology, both human and zoological, encounters when it leaves the broad field of general considerations, to enter upon the more particular study of individual and concrete cases with divergent possibilities of interpretation. For we have direct and immediate acquaintance with no other psychical processes than those which we can study by the introspective method in ourselves. Hence introspective study must inevitably be the basis and foundation of all comparative psychology.

I will endeavour to illustrate the fundamental difficulty of comparative psychology by means of an analogy. Suppose that a chronometer were gifted with intelligence and reason, and were to enter upon the study of other timepieces, all access to their works being inexorably denied it. It would be able to observe the motions of the hands over the dial-plate, and perhaps gain some information by attentively listening to the internal sounds. But when it came to the interpretation of these observed phenomena, and when it attempted to explain their inner causes, the chronometer would be forced to frame all such interpretation and such explanation in terms of its own works. With no other works would it have any acquaintance. It would infer, and justly infer under the circumstances, that the works of other timepieces were, on the whole, of like nature with those which actuated the movements of its own hands over the dial-face. There can be no question, moreover, that the more thorough and accurate the acquaintance of the chronometer with its own works, the more valid would be its inferences with regard to the hidden works of other timepieces. For example, it might learn by introspection that it possessed a mechanism of compensation for changes of temperature; and noticing that in other timepieces the rate of movement of the hands varied with the rise or fall of the thermometer, it might infer that in them such mechanism was absent. It is probable, however, that the chronometer would interpret all the phenomena as due to the action of a mainspring, since it would necessarily be unacquainted with the impelling motive power derived from the descent of heavy weights; and the outflow of energy from the spring would, in its interpretation, be regulated by some sort of balance-wheel, since the principle of the pendulum would nowhere be found through introspection of its works. Thus there would be for the chronometer inevitable possibilities of error. And although it could do little more than speculate concerning these possibilities, it would certainly be wise in refraining from anything like dogmatism concerning the insides of other timepieces which it must interpret, if it interpret at all, in terms of its own chronometer works, but which might not impossibly, could it only get at them, exhibit the application of other mechanical principles.

Now this analogy must certainly not be pressed too far. It is here adduced to illustrate the fact that just as the supposed chronometer would be forced to interpret the mechanism of other timepieces in terms of its own mechanism, so man is forced to interpret the psychology of animals in terms of human psychology, since with this alone he has first-hand acquaintance through the study of the nature and sequence of his own mental processes. But it will perhaps be said that the analogy is invalidated on the principles I have myself adopted, by the fact that in animals a knowledge of the organic mechanism, the functional activity of which is the objective aspect or correlate of psychical processes, is not beyond our reach, but is attainable through physiological research. Access to the works of other timepieces, at any rate from the objective side, is, it may be said, *not* inexorably denied to man the investigating chronometer. So far from this being the case, it is the comparative study of other "works," taken in connection with the comparative study of the life-activities effected through their means, that affords the justification of *inferential* conclusions concerning the psychical processes of animals. This view of the matter, in which I concur, does not seem to me wholly to invalidate the chronometer analogy; but it does suggest a modification, and further development of the analogy.

The chronometer, we will suppose, is acquainted through introspection with its own psychology, and is able to take to pieces the works of other timepieces. It finds a number of chronometers whose works are all practically identical, and as it believes, but cannot demonstrate without taking itself to pieces, just like its own; and it is led to the inference that their psychology is similar to its own. It finds also a number of other timepices whose works are constructed on similar principles, and differ chiefly in their being less highly finished and somewhat less complex; and it is led to the inference that their psychology, though less developed and less complex than its own, has probably been evolved on similar lines. But when it comes to the kitchen clock, it finds certain general similarities, cog-wheels and chains and so forth, but it also discovers principles of construction so different, the weights and the pendulum being so unlike its own balance-wheel and escapement, that it hesitates to draw any positive and definite conclusions. It sees that though the psychology of the kitchen clock may be closely analogous to its own, it may be quite different. It refuses to express a definite opinion on the psychology of the kitchen clock.

To apply the analogy in this modified form. Man, by anatomical and physiological research, has found in other men cerebral hemispheres with sensory-centres, control-centres, and so forth, similar to those which he believes that he individually possesses; and he infers that their psychology is of like nature to his own. He also finds in other vertebrates cerebral hemispheres, with sensory-centres and so forth, differing from man's chiefly in mass and complexity; and he infers that their psychology, though less developed and less complex than his own, has probably been evolved on similar lines. But when he comes to the insect, the crustacean, the mollusc, not to mention the worms, the sea-anemone, or the amœba, he finds nervous systems so different in types of structure from his own, that he hesitates to draw any definite and positive conclusions concerning the psychical states of these animals. It is true that there are nerve-fibres and nerve-cells; but the manner of their arrangement is so different from that of the vertebrates to which he belongs, that the careful student of zoological psychology is forced to conclude, that though the psychical states of insects and crustacea may be similar to those of man, they may be markedly dissimilar.

It may indeed be contended that community of environment—the joint-tenancy of the same world—must necessarily beget community of psychical faculty to meet the requirements of that environment. But while admitting the soundness of this argument so far as it goes, I venture to submit that it does not go far. For why should the community of psychical nature be greater than that of physical nature? The anatomy or the physiology of insects, for example, differs tolerably widely from that of man; why then should he suppose that their *psychical* endowments are more closely similar? Both physical nature and psychical nature are so to speak moulded in accordance with the environment. To both the argument of a joint-tenancy of the world applies. The physical nature being widely divergent from that of man, is it not reasonable to suppose that the psychical nature is, or at least may be, also widely divergent?

No one can study with any attention and care the habits and activities of such insects as ants and bees, without feeling convinced that they profit by experience and that their actions are under control. It is true that at present we know little about the physiology of this control, and of the relation of control centres to automatic centres. But this may sooner or later be remedied by an extension of our knowledge of the nerve-physiology of insects. I am the last to think of counselling any abatement of zeal in the fascinating study of the activities, and of the minute anatomy and physiology of the higher invertebrate forms of life. But I am of opinion that students in this department of investigation may do well to lay to heart the lesson conveyed by the analogy of the chronometer and the kitchen clock. In any case, in an introduction to comparative psychology, I feel bound to lay stress on the necessity for the greatest caution in the psychical interpretation of insect activities; and I feel justified in restricting myself, in this work, to a consideration of the psychical states which we may infer to be associated with the functional activity of the cerebral hemispheres in the higher vertebrates.

Now it is clear that the validity of our inferences concerning the mental processes which underlie the actions of our human neighbour, is primarily dependent on the similarity of his mind to our mind. If he is an Englishman, of the same social grade as ourselves, of like tastes and habits of thought, educated under the same school system, the similarity will be fairly close, though even here there must be slight individual differences. But if he be a foreigner, of a different social grade from ours, differing from us in tastes and habits of thought, educated on other school systems than ours, there will be a wide margin of dissimilarity. We shall find no little difficulty in putting ourselves in his place, in understanding how with such and such facts staring him in the face he can hold the views he says he holds, and in conceiving how he can derive any pleasure from that which would bore us to death, or would set our æsthetic teeth on edge, or would painfully shock our moral sensibility. In dealing with North Australians, or Maori or South Sea Islanders or Red Indians, our difficulties are proportionally increased. These are peoples who have been living, generation after generation, under circumstances widely different from those in which our own race has been nurtured. How difficult it is justly to interpret their thoughts and feelings, and to reach the mental processes which are the psychical accompaniments of their actions, to us so strange and meaningless! And the difficulty is due to the fact, that the only mind with which we can claim any first-hand acquaintance is the civilised mind, that of which we are conscious within ourselves. In the terms of this mind, that of the aboriginal Australian or Red Indian has to be interpreted. We must remember that among civilized men careful introspection and comparative study have led to the formation and adoption of tolerably clear-cut and self-consistent views of the world, and of our relations as individuals to that which we regard as universal. But among primitive folk, of less introspective and reflective power, we must not expect such definiteness and self-consistency. And I confess that I read with some scepticism much that is written on the animistic or fetishistic or other interpretations which are read by philosophers into the vague and often contradictory beliefs of uncivilized peoples. It is difficult for us to realize with what content such peoples can hold a number of beliefs which *to us* appear self-contradictory, and how readily they are satisfied with isolated fragments of explanation, having little or no desire to combine them into a consistent whole. Again, in our very midst there are beings, so like us, and yet so different, the understanding of whose mental processes is difficult and yet most important. I refer to our own children. How unexpected are often the actions of children! How strange their whims and moods and fancies! How charmingly illogical and irrational they sometimes are, and yet often how surprisingly sharp and clever! Notwithstanding the excellent work that has been done in this branch of study, the psychology of the child is a field in which much careful observation and much cautious inference is still needed. And why are the difficulties of interpretation so great? Because we have to interpret in terms of the adult-mind the child-mind, in which the relative development of the faculties, like the relative development of the bodily organs, is so different from that of men and women. It is true that we ourselves have once been children. But what most of us remember of our child-days is not the nature of our mental processes, but certain salient products. A greater or less number of striking external incidents, a few occasions of keen joy or bitter sorrow, constitute for most of us the sum-total of our memories of childhood. To reach mental processes needs introspection, and few children have the power of introspection, or the knowledge by which such introspection must be guided. We have no recollection of the mental processes of childhood, because those mental processes could not then for most of us be objects of thought and contemplation.

Even in human psychology, therefore, if we include not only the psychology of sages, but of ordinary folk, of savages, and of children, there are serious difficulties of interpretation. Since the validity of our inferences concerning the mental processes which underlie the actions of our neighbour is primarily dependent on the similarity of his mind to our mind, it is clear that, if through divergence of development, or imperfection of development, his mind has come to differ from ours, the validity of our inferences will be so far impaired. For we cannot get at his mind directly; our inferences must always be, for better for worse, in terms of our own mental processes.

It will thus be seen that in studying other minds through their objective manifestations, it is primarily essential that we should have, so far as is possible, a thorough and accurate acquaintance with the only mind we can study at first-hand and directly, namely our own. Without this, anything like scientific interpretation is manifestly impossible. All rational human beings have, however, some acquaintance with the workings of their own consciousness. And many of those who are not professed psychologists have, through unusual powers of introspection and keen insight, reached conclusions which are just and true, though they are apt to be somewhat lacking in balance. Psychologists make, or should make, no claim to any monopoly of knowledge in the subject they study; their province is mainly to systematise that knowledge. They bear to the acute and accurate observer the relation of the trained biologist to the amateur naturalist. And just as the amateur naturalist is apt to regard the scientific biologist with some suspicion, as one who is over-subtle, and relies too much on the delicate methods of the laboratory and the dissecting-table; so is the plain man of shrewd insight apt to regard the psychologist also with some suspicion, as one who is over-subtle in his distinctions, too introspective, and not sufficiently objective in his study of mind. And the psychologist should accept the criticism, not with impatience and the assumption of an air of superior knowledge and wisdom, but with a quiet determination to justify his procedure by the results which, through its means, he is enabled to reach. From the position that the first duty of a psychologist is to attain accurate and systematic acquaintance with the workings of his own mind, as the cipher in terms of which all other minds must be read, he cannot recede without abandoning the only basis of scientific method possible under the circumstances of the case.

With this as a basis, he may proceed to the so-called objective study of mind—that is to say, to the study of the objective manifestations in other beings of a consciousness more or less similar to that of which he has, through introspection, some first-hand knowledge; his aim being, by such study, to reach an inferential or second-hand knowledge of the nature of the consciousness which actuates the conduct of these beings. And here the student of human psychology is in a position of great advantage, as compared with that which the student of zoological psychology must rest content with. For, by means of specialized objective manifestations, especially of language, self-conscious human beings can signify to each other the nature of their individual conscious experience. Objective manifestations of some kind are the only index we have of the inner psychical experience. But by means of a common language, human beings can purposely set the index, so as to suggest the particular nature of this psychical experience.

Let us revert once more for a moment to the analogy of the chronometer. In two similar chronometers the position of the index-hands at any moment indicates the exact configuration of the wheels of the internal mechanism. Either of these two, then, which knew from introspection the psychical aspect of its own inner processes, would be able to infer with accuracy the exact configuration of the wheels of its neighbour's works from the position of the index-hands. If it had also a pair of subsidiary hands which it could shift at will, it would be able to indicate at any moment to its neighbour the particular configuration of some previous moment. Thus each could gain inferentially and indirectly through the objective study of the index-hands an accurate knowledge of its neighbour's inner mechanism. Everything, however, would depend upon the similarity of the internal mechanism in the two chronometers. If they were slightly different, they would have very great difficulty in conveying to each other the nature of this difference. So too with human beings. In so far as men are similar in psychical endowments, they can convey to each other through the index of language the nature of their psychical experience. They have great difficulty in making each other acquainted with their individual differences. And the difficulty is the greater the more these individual differences are qualitative, and not merely quantitative. Among civilized men, of like social grade, and somewhat similarly educated, the individual differences are mainly quantitative,—that is to say, differences in the relative development of similar faculties. Careful objective study enables us to gauge and assign a value to the ratio of the faculties in our friends and neighbours. But in the study of uncivilized men, not only of different social grade, but living under a different social system, men whose education and upbringing has been far other than those under which our own character has been moulded, we find differences which are not merely quantitative but qualitative. There is not merely a difference in the ratio of similar faculties, but a greater or less divergence in the nature of the faculties themselves. In such cases our inferences are much more difficult and much less trustworthy.

The realization of the difficulties inseparable from the subject, and of his liability to error in the interpretation of the facts, does not, however, deter the student of human psychology from a careful investigation of the objective manifestations of mind. He takes every opportunity of studying these manifestations, not only in normal men and women of all grades and of all races, not only in normal children and infants, but in pathological cases in hospitals and asylums, and under those abnormal conditions which are presented by patients in the somnambulistic or hypnotic state. Throughout the whole of these objective investigations, the wise and cautious student never forgets that the interpretation of the facts in psychical terms is based upon the inductions he has reached through introspection. The facts are objective phenomena; the interpretation is in terms of subjective experience; and no one has or can have any subjective experience other than that afforded by his own consciousness.

We are now in a position to see clearly what is the distinctive peculiarity of the study of mind in beings other than our own individual selves. Its conclusions are reached not by a singly inductive process, as in Chemistry or Physics, in Astronomy, Geology, Biology, or other purely objective science, but by a doubly inductive process. Inductions reached through the objective study of certain physical manifestations have to be interpreted in terms of inductions reached through the introspective study of mental processes. By induction I mean the observation of facts, the framing of hypotheses to comprise the facts, and the verification of the hypotheses by constant reversion to the touchstone of fact. Our conclusions concerning the mental processes of beings other than our own individual selves are, I repeat, based on a two-fold induction. First the psychologist has to reach, through induction, the laws of mind as revealed to him in his own conscious experience. Here the facts to be studied are facts of consciousness, known at first-hand to him alone among mortals; the hypotheses may logically suggest themselves, in which case they are original so far as the observer himself is concerned, or they may be derived,—that is to say, suggested to the observer by other observers; the verification of the hypotheses is again purely subjective, original or derived theories being submitted to the touchstone of individual experience. This is the one inductive process. The other is more objective. The facts to be observed are external phenomena, physical occurrences in the objective world; the hypotheses again may be either original or derived; the verification is objective, original or derived theories being submitted to the touchstone of observable phenomena.

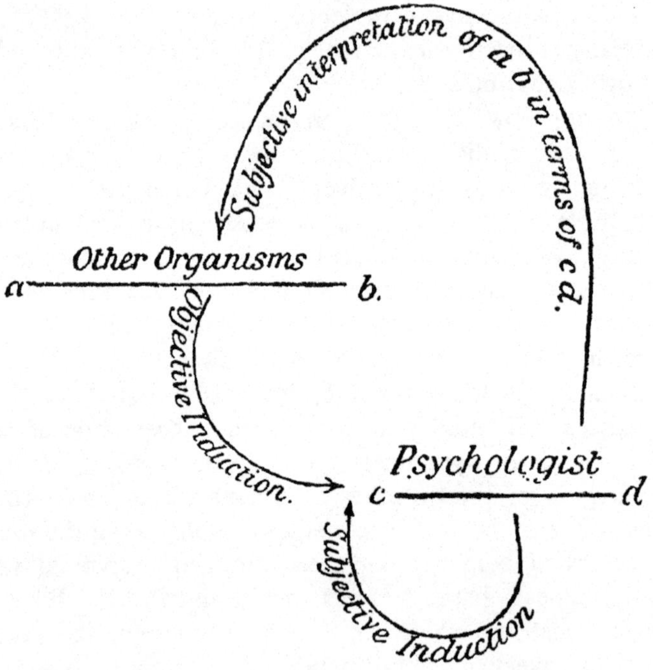

FIG. 7.

Both inductions, subjective and objective, are necessary. Neither can be omitted without renouncing the scientific method. And then finally the objective manifestations in conduct and activity have to be interpreted in terms of subjective experience. The inductions reached by the one method have to be explained in the light of inductions reached by the other method.

I am anxious to make this matter quite clear, and I will therefore endeavour to illustrate it diagrammatically. In the first diagram (Fig. 7) the line *a b* represents the conduct, activities, and other objective phenomena exhibited by other beings or organisms than the individual psychologist, while *c d* represents the states of consciousness of which he alone has direct knowledge.

Then the diagram is intended to show how the psychologist must combine both objective induction and subjective induction, that he may reach a subjective interpretation of *a b* in terms of *c d*.

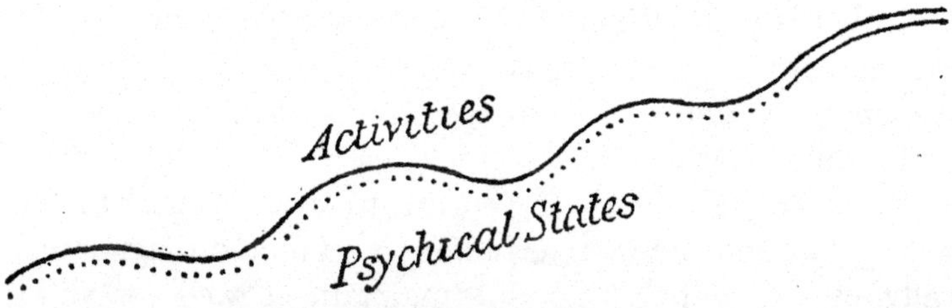

FIG. 8.

In the second diagram (Fig. 8) the curve represents the living beings under investigation comprising the psychologist himself and other organisms. The upper line represents the physical objective aspect, the lower line the psychical subjective aspect. The whole of the upper line is open to the inductive investigation of the psychologist. But of the lower line only the firm part, that representing his own inner consciousness, is open to his inductive study. The dotted part, that representing the mental side of other organisms, has to be interpreted by the inferential prolongation of the known part of the curve, or, in other words, has to be interpreted in terms of the subjective inductions reached through introspection.

Now it is idle to assert that one set of inductions is more important than the other, since both are essential. But there can be no question that the subjective inductions are in some respect more subtle and difficult and delicate than the inductions concerning objective phenomena. There can be no question that false assumptions and vague generalizations more commonly pass muster with regard to mental processes than with regard to their physical manifestations. And there can be no question that in the systematic training of the comparative psychologist the subjective aspect is not *less* important than the objective aspect.

The question now arises whether in passing from human to animal psychology any other method of interpretation is possible than that which holds good for the former. Can the zoological psychologist afford to dispense with that systematic training in introspective or subjective analysis and induction which is absolutely essential for the student of human psychology? I venture to contend that he cannot. The scheme of interpretation exhibited diagrammatically in Fig. 7 holds good I maintain as well for animal psychology as for the psychology of man. There are, I am well aware, many people who fancy that by the objective study of animal life they can pass by direct induction to conclusions concerning the psychical faculties of animals. But this is, I think, through ignorance of the methods of psychology; or perhaps one may say, without injustice, through ignorance of the method that they themselves unconsciously adopt. All that is necessary, these people will tell you, is to observe carefully, and to explain the actions you observe in the most natural manner. "In the most natural manner," here means and is equivalent to, in just the same way as you explain the actions of your human neighbours and acquaintances. And these human actions are explained on the assumption that your neighbour is actuated by motives and impulses similar to your own. Thus these observers who think that their explanations are reached by direct induction are really proceeding unconsciously on the method they affect to disregard. Reduced to its logical basis their contention is that the thorough and systematic study of that mind in terms of which they unconsciously interpret all other minds is unnecessary if not misleading.

Now it appears to me that the foundation of this erroneous view, for as such I must regard it, is the tacit assumption that what suffices for practical purposes suffices also for scientific purposes. All fairly successful men and women acquire, and must acquire, a knowledge of human nature sufficient for the practical needs of everyday life under social conditions. Over-subtlety and refinement of analysis, too great nicety of interpretation, are rather disadvantageous than otherwise in the practical conduct of affairs. Hence practical men are wont to look with some suspicion at the psychologist as one who is prone to be a mere theorist. In the same way practical politicians not uncommonly look with suspicion on sociologists and political economists, practical engineers regard with similar eyes the subtler theories of the physicist, practical metallurgists look askance on the more delicate methods and more advanced hypotheses of the chemist, and in general the practical man is inclined to utilize the results of the man of science but to regard his more refined interpretations of natural phenomena as mere theory.

There can be no question that the interpretation of the actions of animals as the outcome of mental processes essentially similar to those of man amply suffices for practical needs. The farmer, the keeper of a kennel, the cattle-breeder, the gamekeeper, the breaker-in of horses, all the practical men who are employed in the breeding, rearing, and training of animals, and the great number of people who keep animals as pets or in domestic service find a somewhat rough and ready interpretation amply sufficient for their purposes in hand. And not unnaturally they are surprised that the explanation which suffices for them with their wide practical experience is found by the man of science to need serious revision and correction. Often unacquainted with the methods and aims of science in its intellectual aspect as endeavouring to interpret the phenomena of nature, often regarding science as the generally unpaid servant of practical utility, they smile if they do not sneer, at the arrogance of the man of science who tells them that the explanation which is good enough for the practical purposes of daily life is not sufficient for the more subtle and refined purposes of scientific interpretation. Be this as it may, I venture to affirm that whereas the man who has to deal with animals for practical purposes can afford to be ignorant of psychological methods and results, the man who would deal scientifically with the psychical faculties of animals cannot afford to be thus ignorant. For the practical man accuracy of observation and careful induction therefrom are of primary importance, validity of psychological interpretation being for him altogether subsidiary. But for the scientific investigator thorough and accurate knowledge of and training in psychology is of at least co-ordinate importance with accuracy of objective observation.

Unfortunately many able men who are eminently fitted to make and record exact observations on the habits and activities of animals have not undergone the training necessary to enable them to deal with the psychological aspect of the question. The skilled naturalist or biologist is seldom also skilled in psychological analysis. Notwithstanding therefore the admirable and invaluable observations of our great naturalists, we cannot help feeling that their psychological conclusions are hardly on the same level as that reached by their conclusions in the purely biological field.

For in the study of animal psychology as a branch of scientific inquiry, it is necessary that accurate observation, and a sound knowledge of the biological relationships of animals, should go hand in hand with a thorough appreciation of the methods and results of modern psychology. The only fruitful method of procedure is the interpretation of facts observed with due care in the light of sound psychological principles.

What some of these principles are we have considered, or shall consider, in this work. There is one basal principle, however, the brief exposition of which may fitly bring to a close this chapter. It may be thus stated:—*In no case may we interpret an action as the outcome of the exercise of a higher psychical faculty, if it can be interpreted as the outcome of the exercise of one which stands lower in the psychological scale.*

To this principle several objections, none of them however of any real weight, may be raised. First there is the sentimental objection that it is ungenerous to the animal. In dealing with one's fellow-man it is ungenerous to impute to him lower motives for his actions when they may have been dictated by higher motives. Why should we adopt a different course with the poor dumb animal from that which we should adopt with our human neighbour? In the first place, it may be replied, this objection starts by assuming the very point to be proved. The scientific problem is to ascertain the limits of animal psychology. To assume that a given action may be the outcome of the exercise of either a higher or a lower faculty, and that it is more generous to adopt the former alternative, is to assume the existence of the higher faculty, which has to be proved. In the case of our neighbours we have good grounds for knowing that such and such a deed may have been dictated by either a higher or a lower motive. If we had equally good grounds for knowing that the animal was possessed of both higher and lower faculties, the scientific problem would have been solved; and the attribution of the one or the other, in any particular case, would be a purely individual matter of comparatively little general moment. In the second place, this generosity, though eminently desirable in the relations of practical social life, is not precisely the attitude which a critical scientific inquiry demands. Moreover, an ungenerous interpretation of one's neighbour's actions may lead one to express an unjust estimate of his moral character and thus to do him grave social wrong; but an ungenerous interpretation of the faculties of animals can hardly be said to be open to like practical consequences.

A second objection is, that by adopting the principle in question we may be shutting our eyes to the simplest explanation of the phenomena. Is it not simpler to explain the higher activities of animals as the direct outcome of reason or intellectual thought, than to explain them as the complex results of mere intelligence or practical sense-experience? Undoubtedly it may in many cases seem simpler. It is the apparent simplicity of the explanation that leads many people naively to adopt it. But surely the simplicity of an explanation is no necessary criterion of its truth. The explanation of the genesis of the organic world by direct creative fiat, is far simpler than the explanation of its genesis through the indirect method of evolution. The explanation of instinct and early phases of intelligence as due to inherited habit, individually acquired, is undoubtedly simpler than the explanation which Dr Weismann would substitute for it. The formation of the cañon of the Colorado by a sudden rift in the earth's crust, similar to those which opened during the Calabrian earthquakes, is simpler than its formation by the fretting of the stream during long ages under varying meteorological conditions.

In these cases and in many others the simplest explanation is not the one accepted by science. Moreover, the simplicity of the explanation of the phenomena of animal activity as the result of intellectual processes, can only be adopted on the assumption of a correlative complexity in the mental nature of the animal as agent. And to assume this complexity of mental nature on grounds other than those of sound induction, is to depart from the methods of scientific procedure.

But what, it may be asked, is the logical basis upon which this principle is founded? If it be true that the animal mind can only be interpreted in the light of our knowledge of human mind, why should we not use this method of interpretation freely, frankly, and fully? Is there not some contradiction in refusing to do so? For, first, it is contended that we must use the human mind as a key by which to read the brute mind, and then it is contended that this key must be applied with a difference. If we apply the key at all, should we not apply it without reservation?

This criticism might be valid if we were considering the question apart from evolution. Here evolution is postulated. The problem is this: (1) Given a number of divergently ascending grades of organisms, with divergently increasing complexity of organic structure and correlated activities: (2) granted that associated with the increasing organic complexity there is increasing mental or psychical complexity: (3) granted that in man the organic complexity, the complexity of correlated activities, and the associated mental or psychical complexity, has reached the maximum as yet attained: (4) to gauge the psychical level to which any organism has been evolved. As we have already seen, we are forced, as men, to gauge the psychical level of the animal in terms of the only mind of which we have first-hand knowledge, namely the human mind. But how are we to apply the gauge?

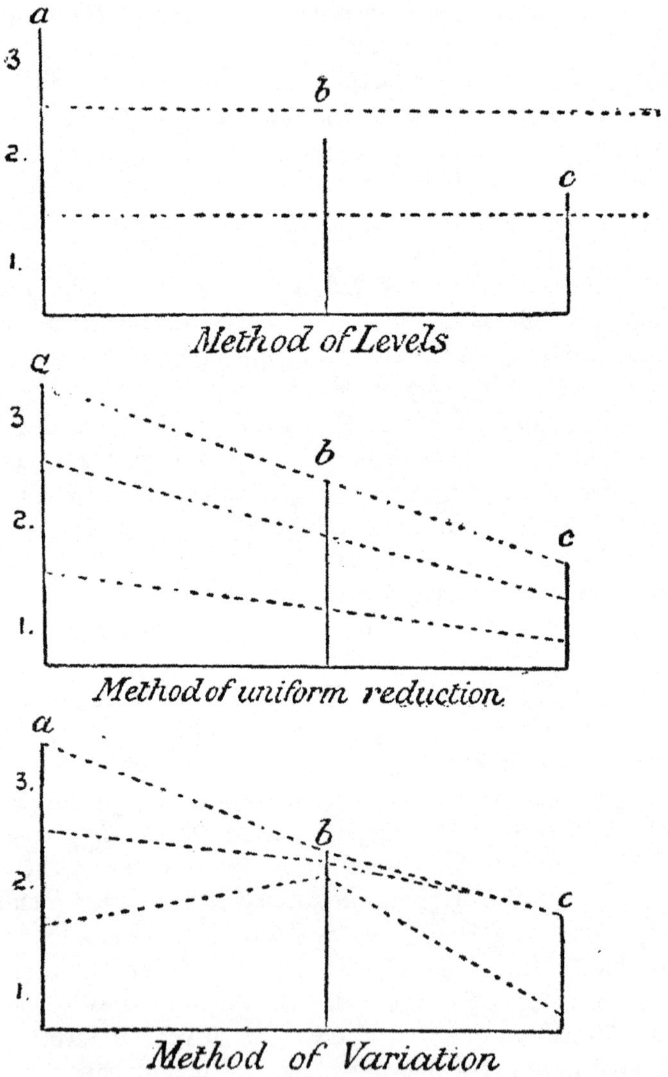
a
3
b
2.
c
1.
Method of Levels
a
3
b
2.
c
1.
Method of uniform reduction
a
3.
b
2.
c
1.
Method of Variation

Fig. 9.

There would appear to be three possible methods, which are exemplified in Fig. 9. Let *a* represent the psychical stature of man, and 1, 2, 3, ascending faculties or stadia in mental development. Let *b c* represent two animals the psychical stature of each of which is to be gauged. It may be gauged first by the "method of levels," according to which the faculties or stadia are of constant value. In the diagram, *b* has not quite reached the level of the beginning of the third or highest faculty, while *c* has only just entered upon the second stadium. Secondly, it may be gauged by the "method of uniform reduction." In both *b* and *c* we have all three faculties represented in the same ratio as in *a*, but all uniformly reduced. And thirdly, it may be gauged by the "method of variation," according to which any one of the faculties 1, 2, or 3, may in *b* and *c* be either increased or reduced relatively to its development in *a*. Let us suppose, for example, that *b* represents the psychical stature of the dog. Then, according to the interpretation on the method of levels, he possesses the lowest faculty (1) in the same degree as man; in the faculty (2) he somewhat falls short of man; while in the highest faculty (3) he is altogether wanting. According to the interpretation on the method of uniform reduction he possesses all the faculties of man but in a reduced degree. And according to the interpretation on the method of variation he excels man in the lowest faculty, while the other two faculties are both reduced but in different degrees. The three "faculties" 1, 2, 3, are not here intended to serve any other purpose than merely to illustrate the three methods of interpretation.

On the principles of evolution we should unquestionably expect that those mental faculties which could give decisive advantage in the struggle for existence would be developed in strict accordance with the divergent conditions of life. Hence it is the third method, which I have termed the method of variation, which we should expect *a priori* to accord most nearly with observed facts. And so far as we can judge from objective observation (the only observation open to us) this would appear to be the case. Presumably there are few observers of animal habit and intelligence who would hesitate in adopting the method of variation as the most probable mode of interpretation. But note that while it is the most probable it is also the most difficult mode of interpretation. According to the method of levels the dog is just like me, without my higher faculties. According to the method of uniform reduction he is just like me, only nowise so highly developed. But according to the method of variation there are many possibilities of error in estimating the amount of such variation. Of the three methods that of variation is the least anthropomorphic, and therefore the most difficult.

In the diagram by which the method of variation is illustrated, the highest faculty 3 is in *c* reduced to zero,—in other words, is absent. It may, however, be objected that this is contrary to the principles of evolution, since the presence of any faculty in higher types involves the germ of this faculty in lower types. This criticism only holds good, however, on the assumption that the evolution of higher faculties out of lower faculties is impossible. Those evolutionists who accept this assumption as valid are logically bound to believe either (1) that all forms of animal life from the amœba upwards have all the faculties of man, only reduced in degree and range, and to interpret all animal psychology on a method of reduction (though not necessarily uniform reduction), or (2) that in the higher forms of life the introduction of the higher faculties has been effected by some means other than that of natural evolution. I am not prepared to accept the assumption as valid; and it will be part of my task in future chapters to consider how the transition from certain lower to certain higher phases of mental development may have been effected.

If this be so it is clear that any animal may be at a stage where certain higher faculties have not yet been evolved from their lower precursors; and hence we are logically bound not to assume the existence of these higher faculties until good reasons shall have been shown for such existence. In other words, we are bound to accept the principle above enunciated: that in no case is an animal activity to be interpreted as the outcome of the exercise of a higher psychical faculty, if it can be fairly interpreted as the outcome of the exercise of one which stands lower in the psychological scale.

CHAPTER IV.

SUGGESTION AND ASSOCIATION.

WE must now return to the wave of consciousness, with a brief consideration of the nature of which we began our investigations. We saw that in any moment of consciousness—any particular *now*—it is experienced as a complex whole, which only on introspection and retrospective analysis can be resolved into its constituents. Let us next consider by what means and in what manner the nature and constitution of the wave is brought about or determined in any series of successive moments of consciousness. In this chapter we shall deal mainly with the wave-crest or focus of consciousness, to the partial neglect of the subconscious body of the wave.

When we take a walk in the country and wisely surrender ourselves to the suggestions of rural nature, the wave crest of consciousness is mainly determined by sights and sounds and scents coming from without. When on our return we sit and read some book with attention, the nature and course of the psychical wave-crest is determined by what the author has written for our instruction or amusement. Or if we listen to our friend describing what he has seen or done in our absence, the course of the wave is guided by his words. In all these cases the determination of the nature of the wave-crest is largely external; but in the first case the external suggestion comes directly from the natural object; while in the latter cases the external suggestion is indirect, through the intervention of symbolic means. For language, written or spoken, is the medium by which we can render the wave of consciousness in our neighbour's mind somewhat similar to that which is passing or has passed through our own.

When, on the other hand, we abandon ourselves to a train of reverie, with closed eyes in a quiet half-hour, or when we are abstractedly thinking out some problem, and are deaf to all that passes around us, then the nature of the wave is not determined from without, but is of internal origin.

But what exactly do we mean when we speak of the determination of the wave as "external" or "internal"? Has the mind, as such, extension or place, so as to justify the use of these physical expressions? Assuredly it has not: and in using the terms "external" and "internal" we have momentarily changed our point of view from that of psychical processes to that of their physical accompaniments. The psychical wave accompanies or is the conscious aspect of the dominant cerebral disturbances in the cortex of the brain. Now the dominance of cerebral disturbances at any moment may be produced in one of two ways, or in part by one and in part by the other. Such dominance may be determined by incoming stimuli through the afferent nerves; this is external determination. Or it may be determined by the preceding cerebral changes, the brain-states of any one moment being the outcome of the brain-states of foregoing moments; this is internal determination. The terms "internal" and "external" are thus, strictly speaking, applicable to the cerebral cortex, which is the seat of molecular disturbances, accompanied by states of consciousness. Where these dominant disturbances are the result of preceding disturbances within the cortex, the determination is internal; where they result from afferent impulses coming from beyond the region of the cortex, the determination is external.

It must be noticed, however, that all the external stimulus does is to set agoing certain internal processes. The impulse from without is like the touch of the trigger which determines a complex explosion. When we are out for a country walk light-waves received by the retina of the eye originate impulses which, conveyed inwards by the afferent nerves, initiate brain-disturbances having for their conscious aspect fields and woods, streams, birds and insects, and so forth. We say that we *see* all these ; and we speak correctly. But in the language of psychological analysis we receive sense-stimuli which suggest all these impressions and certain ideas arising out of them to the mind. Certain notes or sounds suggest the thrush, the cuckoo, or the grasshopper : certain odours, violets or limes.

When we are reading or listening to a speech, states of consciousness are evoked by the printed words (retinal stimuli) or by the sounds which fall on our ear (auditory stimuli). The suggestion of the states of consciousness is external. But the nature of the states depends upon the internal conditions and organization of the brain-centres. We may term the initiation of states of consciousness by external stimuli *primary suggestion*, reserving the phrase *secondary suggestion* for the wholly internal determination of states of consciousness by changes within the cortex.

Taking primary suggestion first, we are not at present in a position to analyse the process very deeply, and must defer its fuller consideration till we come to the chapter on *The Analysis of Impressions*. I shall describe as an *impression* that which is brought to the focus of consciousness, or to the crest of the psychical wave, through primary suggestion. And for the present it must be sufficient to note that the nature of the impression thus suggested by a simple stimulus or by several simple stimuli, is determined (1) by inherited brain-structure or its psychical correlative mental constitution (2) by the modifications impressed on this brain-structure or mental constitution by individual experience. In brief, the existing condition and mode of response of the brain is due in part to inherited structure and in part to individual acquisition. So far as the impressions are concerned, that is to say the focal states suggested by simple afferent stimuli, these *impressions* are presumably very similar or practically identical in all normally constituted human beings. My friend and I have closely similar impressions of the road along which we are walking, of the hedge that borders it, of our dogs racing on in front of us, of the setting sun and the deeply-tinted clouds on the horizon. It is true that no two of us taking the same country walk will have quite the same states of consciousness aroused by external suggestion, because no two of us have had quite the same individual experience, and in no two of us is the mental constitution and inherited brain-structure identical. But the difference here lies not so much in the impressions, or focal states, as in the marginal states which form the body of the psychical wave. Thus, although the *impressions* are closely similar, the *states of consciousness*, which include not only the focal impressions but also their marginal setting, are different.

As my friend and I are walking along the road, during a pause in our conversation we pass a gate at which some cattle are standing. We both begin to speak at once, and, after mutual apologies and the usual courtesies, he takes the precedence, and tells me of the Red Devons with which he has stocked a farm which he has lately purchased. When he has spoken, he asks me what I was about to say; and I laughingly reply that I was merely going to ask whether he thought certain recent promises to electors (1892) were much more likely to be fulfilled than certain other promises in 1885 concerning three acres and a cow. Now here a similar impression, the result of primary suggestion, gives rise in two different minds to two different trains of ideas.

Before going further we must define what is here meant by the word "idea." We used the word "impression" to signify that which is brought to the focus of consciousness, or the crest of the psychical wave, through primary suggestion. We shall use the word "idea" to signify that which is brought to the focus of consciousness through secondary suggestion. The idea may be a clear-cut mental image, like the remembered face of a dear friend, or it may be as vague as that which attaches to the word "universe" or the word "fog." Whether well or ill-defined, it occupies the crest of the psychical wave and has been brought there, not by primary suggestion through the channels of sense, but by secondary suggestion as the result of foregoing brain-disturbances. Thus the primary impression *cow* gives rise in my friend's mind to the secondary idea *farm*, and in my mind to the secondary idea "*elections*." These ideas may suggest others, and so on in a longer or shorter train of secondary suggestion.

It is somewhat difficult to assign exactly the limits which demarcate primary from secondary suggestion. If, for example, I see on Exmoor a footprint which suggests a red-deer; shall we say that the impression is that of a footprint which in turn gives rise to the idea symbolized by the word "red-deer"? or shall we say that an impression of a red-deer is primarily suggested? In this case we may perhaps fairly adopt the former view, and say that the footprint suggests the idea of the beast that made it. But if I see over the brow of the hill the points of an antler, shall we say that the impression of a bit of an antler suggests the idea of the stag that bears it? or shall we say that the visual stimulus from a portion of the stag suggests the impression of the red-deer? Let us suppose we still adopt the former view and say that the impression of an antler-point suggests an idea of the stag. Then how will it be if I see the head and shoulders, while the legs and hind-quarters are hidden? Surely here we may fairly say that we have a direct impression of the red-deer. If so, we imply that a small portion of the stag suggests an impression which gives rise to an idea of the animal itself, while a larger portion directly suggests an impression of the animal. I have no desire to be over-subtle, and I think it probable that if two men—one an experienced and enthusiastic staghunter, the other a mere townsman—were standing together on Exmoor and saw the points of an antler; in the one case *red-deer* would shoot to the focus of consciousness as a practically direct impression; while in the other a vague "something visible" would slowly take form as an impression, to be gradually followed by an idea of the animal, with some such train as this: "Why, it must be an antler; a red-deer, by Jove!" These cases, and a hundred other such, show that the demarcation between primary and secondary suggestion is not very easy to draw with strict rigidity.

It we adopt this interpretation, however, we must admit that the impression may contain a greater or less amount of ideal supplement. Only the antler-tip is presented to sense: the rest is ideally added.

Returning now to the case above noticed, where the same impression *cow* gave rise for me and for my friend to different trains of ideas, there is not much difficulty in assigning, in general terms, reasons for the different results in his mind and in mine. His farm in Devonshire had been for some time a topic of thought and discussion, his mind had a constant tendency to revert to this subject. Presumably from the physiological point of view certain cortical centres, the disturbances in which are associated with this particular form of consciousness, were already in a state of irritability or incipient change, and only needed a suggestive impulse to raise their molecular thrills into dominance. Probably the farm was lurking in the background of his consciousness as he walked silently by my side. On the other hand, my own mind was, as we say, full of the elections, and of certain statements reported to have been made in

Wiltshire to catch the agricultural vote. The cow appeared to me therefore in an electioneering connection. Had a butcher been with us, the cattle might well have suggested the peculiar excellence of last year's Christmas beef. Or if a student of pre-historic archæology had been there, his mind, through the intervention of *Bos primigenius*, might have wandered to the Europe of primitive times.

All this serves to bring into view what appears to be of great importance in serving to determine the direction of the course taken by the psychical wave in secondary suggestion, namely immediate interest. There are certain permanent interests concerning our own future, that of those near and dear to us, that of our own past work, that of our business or occupation, that of our favourite schemes. All these are more or less directly associated with ourselves. In each moment of consciousness they do form or may readily be caused to form part of the body of the wave. They are constantly either in or near the margin of consciousness. There are also the more transitory interests of our immediate thought or occupation, the game we are playing, the book we are reading, the problem we are thinking out. These, too, form important constituents of the psychical wave in the moment in question. If we call the former general interest, and the latter special interest, we may say that the course of the wave, in secondary suggestion, is largely determined by the joint action of general and special interest. Although in a less degree interest has some effect in primary suggestion. Every one knows the story of Eyes and No-eyes. Interest largely makes the difference between the two. If a naturalist is out for a country walk with one whose absorbing pursuit is abstract mathematics, the former will hear and see much to which the latter is psychically deaf and blind. The retina and the auditory nerve-ends are stimulated, but the subconscious effect produced has no chance of rising into dominance when it has to compete with a matter of such absorbing interest as a problem in four-dimensional space.

The course of the psychical wave, then, is largely determined by interest; and the interests of our life are constantly lurking in the margin of consciousness. They form a sort of permanent body to the psychical wave. In saying, therefore, that the wave is largely guided in its course by interest, we are saying in effect that in any series of moments of consciousness *a b c d*, the state of the wave at *c* is determined by the state of the wave at *b*, and will in turn determine the state of the wave at *d*. In other words, the interest is internal, and something the wave carries with it; not external, and moulding the wave of consciousness from without. The only qualification of this general statement we should have on deeper analysis to make, is that probably not only the subconscious states of the body of the wave, but also various infra-conscious or extra-marginal elements below the threshold of consciousness, would have to be taken into consideration. We are burdened, for example, with a great sorrow, but have to fulfil an engagement to lecture or make a speech. Beginning with an effort, we erelong get interested; thoughts connected with our purpose in speaking occupy consciousness; our sorrow for the time is forgotten. We walk away animated perhaps with success; but from out of the unconscious there rises a numb and nameless feeling, and our sorrow regains its sway. It has in the excitement of speaking been thrust below the threshold of consciousness out into the extra-marginal region. But no sooner does the excitement subside, than it rises first into numbing subconsciousness, and then with a pang becomes dominant and focal.

Returning once more to the example we have taken to illustrate secondary suggestion, it is clear that the impression *cow* would never have suggested *farm* to my friend's mind, nor "*elections*" to my mind, in the absence of previous experience. Internal or secondary suggestion does but follow lines previously marked out by external or primary suggestion. A London child visiting the country, and seeing for the first time the yellow flowers of the gorse or furze, is told how Linnæus, the great botanist, when he first saw a furze-bush in golden blossom, knelt down and thanked God that he had been permitted to see such a sight. Certain states of consciousness are thus directly suggested by the sight of the flower and the telling of the incident. Subsequently the sight of the furze-blossom will recall the story of the incident to the child's mind. *The wave of consciousness will tend to repeat its former course.* Not only the story of the incident, but its narrator, and the whole scene, will be more or less clearly suggested. And then the idea of the narrator will perhaps remind the child of another incident she told on the same occasion concerning Charles II. in the oak-tree; and this again will call to mind the history lesson that has to be learnt for to-morrow; and so on.

The wave of consciousness, then, in secondary suggestion tends to repeat its former course, that which was determined by the primary suggestion of direct experience. The occurrence of two impressions in immediate sequence establishes a link between them; and the subsequent occurrence of the first impression suggests the idea which answers to the second. The name given to this link is *association.* The sight of furze-blossom, and the story of Linnæus, are associated in primary experience; and the subsequent sight of yellow furze suggests the story by association. But though the wave of consciousness in secondary suggestion tends to repeat its course, it only does so imperfectly, and for a short distance. It is at any moment liable to be switched off in a new direction, owing to the influence of a stronger association. Thus the sequence in the child's mind when he again sees the yellow furze is,—Linnæus: narrator: oak: English history: to-morrow's lesson. This is not by any means an accurate reproduction of any one previous sequence, but is made up of bits of several previous sequences.

It is clear that in the case of all common objects of which we have had direct experience under varying circumstances, and in the case of many common words which we have used in numerous connections, there is not one particular line of association with each, but rather a number of divergent lines. Indeed, the associations with each familiar sight and sound, and each familiar word, are so numerous and so divergent, that they have, in adult life, little tendency to direct the course of the wave of consciousness in any particular direction. It would be exceedingly inconvenient if they had. If each predicative word in a sentence tended to switch off the wave of consciousness on some particular line of its own, we should lose ourselves in the maze of side issues. If "lose" necessarily reminded us of some particular game, or sum of money we had lost; if "maze" sent us off to wander in memory between thickset hedges at Hampton Court; if "issues" led us in thought to some spring where the water issues from the earth;—how confused would be the result of reading the latter half of the last sentence. And no doubt this is one of the difficulties of early education, and one of the causes of the apparent inattention of children. Experience being limited, the words have particular associations which divert the attention from the subject-matter in hand.

Divergent association is therefore a great gain. For since of the many possible diverging lines there is no inherent tendency to follow any particular one, the mind hangs poised, following none, but ready to follow any of them. But how comes it, if this be so, that in reading a paragraph each word helps to carry the mind—for at each moment of consciousness the existing phase of the psychical wave is for empirical psychology the mind—along the special line intended by the author? If each word has divergent associations, how is the particular association intended by the writer suggested? On the principles already laid down the answer to this question is not far to seek. The word does not stand alone in the paragraph, and that which it suggests is only part of the psychical wave. The moment of consciousness embraces not only the particular word in the focus, but a few words that have not reached the focus, and half a sentence or more that has passed the focus and is fading out of marginal consciousness. And since the psychical wave in succeeding moments of consciousness is continuous, it is determined at any one moment by what it has been at previous moments. When we are reading any particular word, what it suggests is determined by the net result of all the preceding sentences in the paragraph.

In like manner, if in a chain of reverie any particular idea or image occurs, what it will serve to suggest depends upon the nature of the psychical wave as a whole, of which this image or idea, though for the moment in the focus, is only one constituent element out of several or many. Much confusion arises, and many unnecessary difficulties present themselves, if we fall into the analyst's fallacy and suppose that the particular idea or image, which we can separate from the rest by retrospective analysis, was separate and distinct from the rest in its natural occurrence. Therefore I again insist on the fact that in the moment of consciousness the psychical wave has numerous elements; and that a state of consciousness is something exceedingly complex, and very different from an isolated image or idea.

So complex is each state of consciousness—that is to say, the psychical wave in each moment of consciousness—that the same state of consciousness probably rarely occurs twice in the experience of any individual. Even when the central impression or idea is the same, its setting, the marginal region or sub-conscious body of the wave, is different. Our surroundings are not quite the same. We ourselves change from day to day and from year to year. The continuous individuality—that felt bond of continuity in conscious existence—is, be it never so slightly, modified by all our experiences. We feel that we are not what we were. The psychical wave at any moment of consciousness leaves traces or habitudes which modify all future phases of the wave. If then a state of consciousness is so complex, and if it is rare that any two states of consciousness should be quite the same, it might seem a hopeless task to endeavour to formulate the laws which condition their being. But here, as in other matters of science, we employ the method of analysis, and, by studying severally, through retrospection, the constituents of the psychical wave, obtain some insight into its constitution and the manner of its origin.

As the result of such analysis certain laws of association have been formulated. As generally accepted they are two —(1) association by contiguity, and (2) association by similarity; to which, as subsidiary, is sometimes added a third, (3) association by contrast.

The law of contiguity may, in accordance with our method of presenting the subject, be thus enunciated:—If any two focal states a and l occur in successive moments of consciousness as impressions, the subsequent recurrence of a as impression or idea will tend, under similar marginal conditions, to suggest the recurrence of l as an idea.

In primitive experience the focal states a and l are determined by primary suggestion, and both are therefore impressions. Thus a vivid flash of lightning is followed in normal experience by a thunder-clap. In the absence of experience there is nothing in the former to suggest the occurrence of the latter. Subsequently the impression of lightning will be followed by an idea of a thunder-clap, and this idea will very probably precede the actual impression, constituting what is sometimes called a pre-perception. Eventually the idea of a flash of lightning, however produced, will be succeeded by an idea of the thunder-clap. A lady of my acquaintance once told me how, after a storm, she showed her little boy a spirited picture of a storm at sea with a ship being struck by lightning. He gazed intently for a few moments and then asked, "Mother, why doesn't it rumble?" The law of contiguity, be it noted, says nothing about the primitive sequence *a l;* that is given in experience. The sequence may be a natural or an arbitrary one; with this the law of association has no concern. It only asserts that if *a* be followed in experience by *l*, then subsequently the recurrence of *a*, either as impression or idea, will be followed by the recurrence of *l* as an idea.

It may be well to note here the limiting conditions introduced in the foregoing enunciation of the law of contiguity. The recurrence of *a* will be followed by the recurrence of *l* *under similar marginal conditions.* We have already seen how the impression *cow* suggested *farm*, &c., to my friend, and "*elections*," &c., to me. The marginal conditions, the settings, so to speak, of the impression, were different in the two cases, and the secondary suggestion was therefore different. Only under similar marginal conditions will the impression *a* suggest *l*. Under different marginal conditions, it may suggest *p* or *z*. It is true that the marginal conditions never or very rarely are quite similar. But such an assumption in the enunciation of a general law is very common in scientific method. The first law of motion asserts that the motion of a body when left to itself will be uniform and rectilinear. Practically no physical body is left to itself; it is never free from the influence of surroundings. But no one with an adequate grasp of scientific method regards this as a defect in Newton's law. Nor should we regard the assumption of similar marginal conditions as any defect in the law of contiguity.

It must not be supposed, however, that the law of contiguity is on the same plane of universality as the first law of motion. There is another implied but unexpressed assumption that is never more than partially true. It is assumed that memory is perfect. That this assumption is false, or only partially true, is only too apparent. We may to-day establish the sequence *iter itineris*, or *seven-times-eight fifty-six*, in a small boy's mind, so that at the end of the lesson *iter* at once suggests *itineris*, and the multiplication-table runs smoothly. But next week *iter* unfortunately does not suggest *itineris*, and the multiplication-table is decidedly lumpy. Those of us who in adult life have weak or treacherous memories, know well that associations are often not permanent but transitory. The wave of consciousness that flowed so glibly from *a* to *l* last week, now flows to any other letter of the alphabet. The bond between *a* and *l* seems to have dissolved. It would be confusing, however, to interpolate in the midst of a consideration of association more than a passing allusion to the conditions of memory. It is therefore sufficient to note here that the relative permanence or transiency of association-links is primarily dependent on unalterable physiological conditions. Macaulay could not help remembering; most of us cannot help forgetting. Secondly, with a given faculty, the relative permanence depends (1) on the strength of the initial association depending largely on the interest it arouses; (2) on the recency of its establishment; and (3) on the frequency of its occurrence. For the present, however, we neglect all these considerations and assume a perfect memory, though, as we shall hereafter see, forgetfulness is one of the conditions of our mental life, and a *sine qua non* of its development.

We are still very far from anything like adequate or exact knowledge of the physiological conditions of association. An external stimulus *x*, giving rise to the impression *a*, is followed by a quite different and unconnected stimulus *g* giving rise to the impression *l*. At first sight there is no very obvious reason why any connection should be established between *a* and *l*. And if the wave of consciousness were discontinuous, there probably would be no connection. But the wave of consciousness is not discontinuous, but continuous. There is not an impression *a*, and then a blank followed by another impression *l*; but the impression *a* passes continuously and without break into the impression *l*. The effects of succeeding stimuli may be represented diagrammatically thus:—

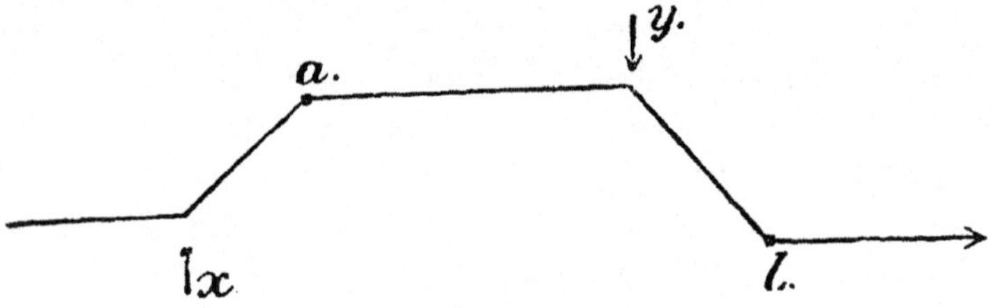

FIG. 10.

The stream of consciousness is represented as flowing continuously from left to right along the line. Its course is deflected by the external stimulus *x*, as the result of which the substantive impression *a* is produced. The stimulus *x* is followed by the different stimulus *y*, as the result of which the wave of consciousness is again deflected, and the substantive impression *l* is produced. Of course this puts the matter in a crude and very diagrammatic way. But it serves to illustrate the hypothesis that the connection between successive impressions is *the transitional state of the wave* between *a* and *l*. Physiologically very little is known of the nervous conditions of the transition. I shall have more to say hereafter on its psychological nature.

To return to the law of contiguity. We have seen that, as above enunciated, it asserts that if any two focal states *a* and *l* occur in successive moments of consciousness, the subsequent recurrence of *a* will tend, under similar marginal conditions, to suggest the recurrence of *l*. As it is frequently stated, however, the law of contiguity seems to assert something more than this. It asserts that either of two ideas associated by contiguity tends to suggest the other. And this as a practical fact is sufficiently conformable with experience. I notice, for example, in my country walk an unfamiliar shrub, and then its peculiar lanceolate leaf. Afterwards the sight of the shrub suggests the leaf, and conversely the sight of the leaf will suggest the shrub.

Let us note in passing, that although the leaf is part of the shrub as a natural object, yet the leaf and the shrub are quite distinct as impressions or ideas. When I have an impression of the shrub, it occupies the focus of consciousness as a whole; but when I concentrate my attention upon one particular leaf, the shrub, though still present to consciousness, becomes merely marginal. I may have an impression (or an idea) of a flock of sheep, of one sheep of the flock, of a portion of the fleece of one sheep, or of one hair of the fleecy wool. All of these are quite separate ideas, which may occupy the focus of consciousness in successive moments, and are capable of becoming associated by contiguity. As the bush may suggest the leaf, so may the flock suggest the individual sheep, and in each case it would seem that the order is reversible. It would seem then that we must modify our previous statement, and assert that the psychical wave may either repeat its previous course, *a* being followed by *l*, or reverse its course, *l* being followed by *a*.

I question, however, whether the wave ever does reverse its course. The story is told of a schoolmaster who asked a sailor lad to box the compass backwards. He readily did so; and then asked the schoolmaster to say with equal rapidity the alphabet backwards. He could not. For the sailor lad, the wave of consciousness had been wont to flow round the compass in either direction indifferently; for the schoolmaster, the wave of consciousness had always flowed in one direction, and he could not reverse its course.

In the many cases where either of two ideas will indifferently suggest the other, as in the case of the shrub and its leaf, it is probable that there has been an alternate occupancy of the focus of consciousness. Shrub, leaf, shrub again, leaf again, and so on. Either will then suggest the other, the wave repeating a part of its former course without reversal. In any case it does not seem in accordance with experience to say that the order of suggestion is in all cases indifferent. Many of us who have a fair reading acquaintance with French or German, speaking it or writing it but seldom, find that though the foreign word at once suggests its English equivalent, the English word does not at once suggest its French or German equivalent. But if we have a conversational knowledge of the language as well as a reading knowledge, this difference disappears. The order of association is here—in my own case at all events—not indifferent. I think we may therefore leave the enunciation of the law of contiguity as above given, requesting those who are satisfied that reversal is also true to modify the statement of the law in accordance with this view.

Passing now to the law of *association by similarity*, we find that this is not so definite and clear-cut, or so general in its application, as that we have already considered. It may be thus stated:—If two focal states of consciousness resemble each other, the recurrence of the one as impression or idea will tend to suggest the recurrence of the other as an idea. Note the ambiguity of the word "resemble." If we ask: Resemble in what respect? we can only reply: Resemble sufficiently in any respect. But we shall find it hard to render definite the sufficiency of resemblance. We may express the law somewhat differently thus:—If two focal states have common elements, the recurrence of one of them will suggest the recurrence of the other. But we cannot define very clearly what either the nature or the amount of the common elements must be to ensure recurrence under the law of similarity.

Standing by the medallion of Wordsworth in Grasmere church, my companion said to me, "It reminds me of Dante." Something about the face, perhaps the nose, suggested that of Dante. When associations of this type occur, we generally express them by some such phrase as "This reminds me of that," or "This resembles that." The crocodile reminds us of an alligator; the slowworm resembles a snake. Instances of such association are so common and familiar that they need no further illustration.

I shall employ the phrase *association by resemblance* for such cases as these where an impression suggests the idea of that which more or less closely resembles it, and shall use the phrase *association by perceived similarity of relationship* for certain more complex cases of far wider range and of a far subtler nature. These cases I will now proceed to illustrate.

Where Shelley says, in his "Ode to the West Wind":—

> "Drive my dead thoughts over the universe
> Like withered leaves, to quicken a new birth,—

the similarity between the focal idea "dead thoughts" and the focal idea "withered leaves" is by no means simple and obvious. One may say indeed that there is no similarity between the focal ideas taken by themselves; they are only assimilated through the imaginative insight of the poet. The metaphors and imagery of every great poet are full of such subtle associations. Take, for example, Shakespeare's

> "Now is the winter of our discontent
> Made glorious summer by this sun of York."

Keats speaks of wine in a goblet,

"With beaded bubbles winking at the brim."

Tennyson, in "Akbar's Dream," writes:—

"And morn
Has lifted the dark eyelash of the Night
From off the rosy cheek of waking Day."

The slow prosaic mind cannot follow these subtle associations, involving delicate perceptions of similarity of relationship, that abound in and give character to the best imaginative poetry. It wonders why Shakespeare drags in a reference to the season of the year, and thinks he should have spelt *sun* of York with an *o*. And when Keats, with his subtle and imaginative play of thought, says that Porphyro

"From the closet crept,
Noiseless as fear in a wide wilderness;"

or when Browning, who had so thoroughly assimilated certain salient scientific truths, says:—

"And I know not if, save in this, such gift be allowed to man
That out of three sounds he frame, not a fourth sound, but a star;"

many complain of their similes as far-fetched. And probably no one who had not a permeating knowledge of the fact that the white light of the star is resolvable into differently coloured rays, would catch Browning's thought that just as different colours combine to form white light, symbolized concretely in "a star," so do different sounds coalesce to form that new and diverse experience a musical chord.*

Wit and humour largely depend upon association by similarity, as where Oliver Wendell Holmes writes, the street-band having ceased its torture:—

"And silence, like a poultice, came
To heal the blows of sound."

* A lady who knows her Browning tells me that she had regarded the "star" as symbolic of the music of the spheres. Perhaps this was also in Browning's mind, a mind ever rich in many-sided allusions.

An able fellow-student of mine, a Frenchman, once complained that he could not follow a train of reasoning given in the morning's lecture. A feather-head standing near exclaimed, "Never mind, old man, we're all rowing in the same boat." "But, thank the Lord, not with the same *sculls*," was the prompt and witty reply. It is Mr Winwood Reade, I think, who somewhere says, "Life is bottled sunshine; and Death the silent-footed butler who draws the cork."

It must now be more definitely noted that these subtler scintillations of a brilliant intellect, though based on similarity, involve something more than simple resemblance of two focal ideas. There is no inherent similarity between silence and a poultice, or between a musical chord and a star. *The similarity is not in the focal ideas, but in their relations.* They may be expressed in the form of ratios. As a poultice is to the physical effects of a blow, so is silence to the effects of a street-band. As coloured rays are to starlight, so are separate notes to a musical chord. As withered leaves form a fertile soil for new plants, so will my dead thoughts form a fertile soil for new thoughts. And similarly in other cases. What we term the relation, is the transition from one focal state to another focal state. In view of these more complex cases of what I have termed association by perceived similarity of relationship, we may add to our former statement of the law the following:—If the relation of two focal states a and l resembles the relation of two other focal states b and m, the recurrence of $a : l$ will suggest the recurrence of $b : m$. Or, more briefly, similar ratios or transitions suggest one another. We cannot follow this up further now. We shall hereafter see that this, which is the basis of poetical imagery, is also the basis of scientific insight and inference.

Concerning the principle of *association by contrast*, little need be said. It is based upon the fact that strongly contrasted ideas tend to suggest each other. A giant suggests a dwarf; beauty, ugliness; virtue, vice; and so forth. In such cases, however, there would seem to be always a bond either of contiguity (*i.e.* previous association in experience) or of similarity. We have perhaps seen or probably read of giants and dwarfs in connection with each other; and they are alike in being abnormal departures from the normal stature of man. That is to say, there is a similarity of relation. Good, bad; beautiful, ugly; virtuous, vicious; great, small; these are terms which, so to speak, go in couples, and thus come to be associated in the experiences derived from ordinary speech. Moreover, contrast is constantly used in early education. "Be good, and not naughty." "This is red and not blue." "What is the difference between solid and liquid?" Thus associations by contiguity are formed, and the perception of similar or dissimilar relations is fostered. Although, therefore, there is no denying the fact that an idea may suggest its antithetical idea, there seems no reason for invoking any principle of association other than those of contiguity or similarity.

Of these two, that of contiguity must undoubtedly be regarded as the more primitive, and it is capable of the more definite and rigid enunciation. Both deal expressly with the focal elements of consciousness. But it must not be forgotten that in normal human experience the focal elements can never be regarded as isolated; and that in any normal sequence of focal elements $m\ n\ o\ p\ q$, the nature of any one of them p is not determined only by the nature of its antecedent o, but also by the subconscious, and perhaps also the infra-conscious, accompaniments of o. This we have already illustrated in indicating the nature of divergent association, and in showing that the particular association in any special case is determined by the general drift of the mind at the time being. I pass in my walk two men in conversation, and overhear the remark, "I don't care much for *bowls*." Owing to divergent associations with the word "*bowls*," this does not convey much to me. But there was probably no doubt as to the meaning in the mind of the person to whom the remark was addressed. Perhaps they were discussing games, or had been talking of the news of the Spanish Armada reaching England; perhaps they were naturalists, and were comparing views as to the best shape for an aquarium tank; perhaps they were talking about Coleridge, and the influence of the sonnets of Mr W. L. Bowles on the poet's mind. I cannot say. It all depends upon the nature of the wave of consciousness into which *bowls* was thus introduced. In autumn, and at evening, we look upon the world with different eyes from those which looked on the spring of the year in early morn. "There is," as Wordsworth says, "an imaginative influence in the voice of the cuckoo, when the voice has taken possession of a deep mountain valley, very different from anything which can be excited by the same sound in a flat country." "After all, it is upon the mind which a traveller brings along with him that his acquisitions, whether of pleasure or profit, must principally depend."

This leads us to notice in passing the influence of the mood one is in, and the temperament one is of, upon the dominant idea of the psychical wave. Mood and temperament express respectively the transitory or relatively permanent condition of the body of the wave of which the dominant idea is the crest or summit. The sequence of ideas is different, according as one is sanguine or melancholic, choleric or phlegmatic. The poetic temperament is in many respects different from the scientific, and both from the merely prosaic. In the prosaic person, association by contiguity reigns supreme; in the poet, association by similarity is rich and abundant. The man of science, if he have the gift of imagination, utilizes association by similarity, but always seeks to make it subservient to, or at least to correlate it with, association by contiguity.

In conclusion, we may ask whether any given state of consciousness regarded as a whole is completely determined by the preceding state of consciousness. That is, using the letters *m n o p q* not merely to represent the focal elements, but states of consciousness in their entirety, is the state *p* completely determined by the state *o*? At a dinner-party the other night I had the misfortune on entering the room to say to my hostess one of those things which are better put differently if not left unsaid. This left a sore place in my mind, which continually smarted. I lapsed into unconsciousness of it from time to time when I became interested in the conversation of my charming companion. But in the pauses of the conversation, back it came into consciousness, and reminded me like conscience of my evil deed. Now here we must either say that during conversation my *faux pas* was present to consciousness, though I was not conscious of it, which sounds to say the least of it somewhat contradictory; or we must say that the state of consciousness *p* was not completely determined by the preceding state of consciousness *o*. The facts are best explained by saying that during conversation the *faux pas* remained in the extra-marginal infra-conscious region, overmastered by more powerful and insistent states of consciousness. If this be so, there are three ways in which a focal element in consciousness may be normally determined:—

(1) In association with cerebral disturbances determined by an external stimulus in primary suggestion.

(2) In association with cerebral disturbances determined by preceding disturbances, dominant or sub-dominant, and therefore associated with consciousness or subconsciousness.

(3) In association with cerebral disturbances determined by preceding infra-dominant disturbances not associated

with consciousness, or whose psychical concomitants are infra-conscious.

According to the doctrine of determinism, every normal * state of consciousness is determined in one or more of these three ways.

---

* Reference to abnormal determination either by a poisoned blood-supply, physical injury, disease, or "thought transference," is here intentionally omitted.

CHAPTER V.

ASSOCIATION OF IDEAS IN ANIMALS.

IT can scarcely be doubted that for animals, as for man, there is a wave of consciousness. By animals I here mean vertebrates, to which, as I have before said, I propose to limit my attention in this Introduction to Comparative Psychology. In mammals and birds, and to a less degree in reptiles, amphibians, and fishes, there is a community of sensory endowment and a community of brain-structure. All these forms of life belong to one great branch of the animal kingdom; and if we accept evolution as the true basis of explanation alike in biology and in psychology, we are justified in inferring that, co-ordinate and concurrent with the community of nervous mechanism and its physiological functioning, there is a community of psychical nature and psychological functioning.

When we watch a cat stealing upon a bird we may fairly conclude that the impression of the bird is focal to consciousness, and that the bird is set in a visual scene the surrounding details of which are merely marginal in the cat's consciousness, as are also the movements of her own limbs, which are being subconsciously controlled in accordance with the nature of the ground over which she is passing. Let us not quarrel over the word "consciousness." There are some who contend that in strict accuracy, that is to say, according to their definition of the word, we must distinguish between the consciousness of man and the consentience of animals. I do not propose to discuss the advisability or the reverse of such limitation of the term "consciousness." I shall in this work use the word quite broadly and generally, so as to comprise both consciousness in the more restricted sense and consentience. I repeat that when the cat is stealing upon the bird we have good grounds for inferring that there are both focal and marginal elements in consciousness; and hence that the state of consciousness at the moment in question might, were our means of acquiring the necessary knowledge less inadequate than they inevitably are, be represented in the form of a curve.

I shall take it for granted then, without further discussion, that in the animals we are considering there are, as in man, both focal and marginal elements in consciousness. The wave of consciousness may be in them far simpler in constitution than it is in man; but I shall adopt the hypothesis that there is such a wave, and shall use the terms "impression" and "idea" for the focal constituents of the wave. Thus the kitten has an impression of the ball with which it is playing, and the hungry dog may have an idea of a nice meaty bone.

That the impression is brought to the focus by primary suggestion through afferent fibres, no one is likely to question. But that the idea is due to secondary suggestion through association it may be well to illustrate, though it is scarcely probable that many would be disposed to doubt the fact.

I have made a series of observations on young chicks and ducks hatched out in an incubator, with the object of studying experimentally the establishment of associations. A few extracts from my note-book will suffice to show the nature of the evidence. A chick, about eighteen hours old, pecked at its own excrement rapidly thrice in succession, and then shook its head and wiped its bill on the ground. Ten minutes later it began to peck, but checked the action before reaching the excrement and wiped its bill. A little later it came near, looked at the material, and then walked away. A visual impression and a taste impression had become associated, and the recurrence of the former suggested a representation of the latter as an idea. On the morning of their second day of life I placed a shallow tin of water before my chicks. They took no notice, and several ran through it without heeding it. Presently one of the birds, while standing in the water, chanced to peck at its toes, as chicks so frequently do. At once he lifted his head and drank repeatedly. Another was led to drink by pecking at a bubble on the brim. Others seemed to do so by imitation of their neighbours. Some time afterwards one chanced to run through the water. It at once stopped and drank. It seemed as if the wetting of the feet suggested the act of drinking—the two experiences having become associated by contiguity. One of my chicks three or four days old snapped up a hive-bee and ran off with it. Then he dropped it, shook his head much and often, and wiped his bill repeatedly. I do not think he had been stung, probably he tasted the poison. In any case, in a few minutes he seemed quite happy and eager after new experiences. But though he came and looked at it once or twice, he made no further attempt to run off with the hive-bee. An association, based on a single experience, was at least temporarily established. Similar experiments with the unpleasant caterpillar of the cinnabar moth and with lady-birds showed that the association between a peculiar appearance and nasty taste was in all cases very rapidly established, and that the visual impression suggested the idea or re-presentation of unpleasant gustatory experience.

A word of warning may here be introduced against a not-unnatural tendency begotten by our living in what may be termed an atmosphere of human conceptions. We call the cinnabar caterpillar an "object," and we say that the object has a certain visual appearance, banded with gold and black, and also, for the chick which takes it into its bill, a certain taste. And when we see that in the chick an association is established between appearance and taste, we are apt, without further thought, to suppose that the chick, two or three days old, distinguishes between the object, its appearance, and its taste, and associates the appearance of the object with its taste as distinguishable and already distinguished qualities. Now I shall have to say somewhat concerning "objects" and "qualities" in the proper place. Here at present we have nothing to do with them save resolutely to exclude them from our thought. The visual impression of the banded black and yellow caterpillar is a bit of simple and direct experience: the gustatory impression which follows is no less a bit of simple and direct experience. And when, subsequently, the sight of the caterpillar suggests its taste, what we have is the presentative occurrence of the one bit of experience suggesting the representative occurrence of the other bit of experience. The little chick does not bother its head in the most rudimentary way about either objects or qualities. Sense-experience is all-sufficient for the practical needs of its simple life.

One of the greatest difficulties against which the student of zoological psychology has to contend is, that the language in which he needs must describe and endeavour to explain the mental processes of animals embodies the results of a vast amount of analytic thought. He has to employ phrases which imply analysis, to describe experiences which involve no analysis. We speak, for example, of the "taste of the caterpillar," which seems to imply the distinction between taste as a property of the caterpillar, and the caterpillar as possessed of this property. For those critics who delight to catch an author tripping in his words rather than in his thought, nothing would be easier than to score a point by exclaiming, "The very language the author employs betrays the fallacy of his contention; he pretends to believe that animals are incapable of analysis; and yet there is scarcely a phrase descriptive of the mental processes of animals which does not plainly indicate such power of analysis." I am desirous that the reader should quite clearly grasp (1) that our language is full of the results of the analysis of phenomena; (2) that this analysis has to be effected by man, the language maker and the language user; (3) that language, being thus saturated with the results of analysis, it is practically impossible to describe mental processes in their primitive unanalysed modes of occurrence without using phrases which are analytic in form; and (4) that the use of such analytic phrases must not be taken to imply analysis in the animal. What we call the "taste of the caterpillar" is for the chick a bit of simple, direct, unanalysed, sensory experience.

To return now to my experimental poultry-yard. I have been much struck, as I watched the progress day by day of my families of chicks or ducklings, with the fact that although they bring with them into the world an inherited aptitude to perform certain activities, yet all experience, even of the performances of these activities, is a matter of individual acquisition. And further, that this experience is rendered of practical value through association. Only in so far as associations are formed does experience afford a basis for the guidance of action, and the conduct of the business of life. It is only as associations are established between impressions of sight and taste that the chick begins to learn what to eat and what to avoid. At first he picks up anything of convenient size that catches his observant eye. Every minute of the early hours and days of life he is establishing associations of eminently practical value for his life's guidance. The environment is simple, and the associations direct and oft-recurring. Hence at the end of a week or ten days he is a remarkably wide-awake little bird.

But although there is, so far as I have been able to observe, no satisfactory evidence of anything like inherited experience, the associations being in all cases individually acquired, I think it extremely probable that there are inherited facilities for association, if I may so phrase it. I mean that there very likely exist in the cerebral hemispheres nerve-tracks which facilitate the establishment of such associations as those between sight and taste. What is inherited, however, is the mechanism by which *an* association may be established; what is a matter of individual acquisition is *the* association that is established. My chicks thoroughly enjoyed the warmth of my hand, and when I stretched it out towards them they would run to it, and nestle in the half-closed palm. I instance this as an association which in this particular form could not have been inherited, and must have been of purely individual establishment.

I may here mention a somewhat amusing case of association in two of my ducklings. I had placed before them regularly at the same time each morning a black tray, on which was a shallow tin containing water; they ran to it, drinking eagerly, sitting in the tin and washing. On the sixth morning I gave them at the usual time the black tray and the shallow tin, but without the water. They ran to it, scooped with their bills along the bottom of the empty tin, and made all the motions of the beak characteristic of drinking. They sat in the empty tin, waggling their little tails and ducking their heads, throwing wholly imaginary water over their backs. This they continued to do for ten minutes,—the action becoming less and less vigorous. Then I gave them water. Next morning I repeated the same experiment. They ran to the tin, squatted in it, and tried to find water to drink; but after three minutes or so went off, but ran to it again when I poured in water. The following morning, after just searching for water in the empty tin, they waddled off.

It must be remembered that the chicks and ducklings upon which my observations and experiments were carried out were entirely dependent upon their own individual efforts in gaining acquaintance with their little world; for they were hatched out in an incubator, and had never known a mother. Under more normal conditions they would have had some guidance from the hen as mother or as foster-parent, and they would have grown up among all the traditions of the poultry-yard. They would therefore have had some guidance in the formation of associations,—imitation of the hen, and to a less degree of other individuals of their kind, aiding them not a little in the acquisition of experience. I was, however, unable to observe, when I returned my young birds to the yard whence I had obtained the eggs, that, at ages of from ten days to three weeks, they were in any marked respect inferior to other chicks of the same age. No doubt there were minor matters in which their experience was deficient; but these were nowise obvious.

It would seem then difficult to overestimate the importance of the association of ideas in the early life of animals. It is the means—the sole means—by which experience is made available for the guidance of action. Experience, no matter how often repeated, is useless for guidance unless associations are established. If a chick takes a ladybird in its beak forty times, and each time finds it nasty, this is of no practical value to the bird unless the sight of the insect suggests the nasty taste before the insect is taken into the bill. And it must be remembered that the utmost which parental care can effect is a certain amount of guidance in the establishment of associations. Association links cannot be formed vicariously. Hence no matter how complete and how long-continued the fosterage in the early days of life, all the associations must be established by the individual himself for his own individual guidance.

I have utilised my observations on young chicks to bring out this conclusion, because in them the establishment of association-links can be so readily watched. But analogous facts can be observed in all young animals. And it is of course those associations which are liable to reiterated re-establishment, when the linkage tends to fade, as it is apt to do in animals and men, that become so ingrained in experience as to be practically indissoluble. A merely chance association may last for a while, but will fade more or less quickly. I was experimenting with a fox-terrier, and testing his intelligence in bringing a stick through railings, narrow gaps in a hedge, and so forth. He was keenly eager to fetch the stick when I threw it just the other side of a hedge. To reach the gap he had to pass through some nettles. By these he was stung, and dropping the stick ran off and rolled in the grass. I fetched the stick, called and patted the dog, and then threw the stick out into the open field. He rushed off; but when he saw the stick on the ground nothing would induce him to touch it. I broke another stick, and he fetched it again and again. But whenever I threw that stick, he refused to pick it up even when coaxed to do so. A temporary association between that stick and the stinging had seemingly been established; but it was only temporary. I took him on for a five or six mile walk, and then returned, after a couple of hours or so, to the same field. I threw the stick; he bounded after it, and fetched it without hesitation. The smart of the nettle had subsided, and with it, seemingly, the association.

I may here give another example of association in the same dog. He will fetch sticks for me by the hour; but if I throw a golf-ball, or even a rounded stone, then sticks at once lose their interest. I may pocket the ball and throw sticks, but the most he will do is run a few yards after them, and then come and beg for the ball. I therefore tried the following experiment:—I placed a ball in my pocket—an old one that he had gnawed almost beyond recognition—and started forth for my walk, repeatedly throwing a stick for him to his perfect satisfaction. I then put my hand in my pocket, fumbled the ball, and patted the dog's head with that hand. He sniffed at it eagerly. I threw a stick; he bounded off a few yards, and then came jumping at my side. A race after the ball had been suggested through the channel of olfactory sensations, but the stick no longer had the same interest.

I think it possible that association through olfactory channels may help us to account for the results of certain experiments I have made with dogs. The experiments themselves were suggested by that recorded by Mr Romanes, in an article entitled "Fetishism in Animals."* Soap-bubbles are blown, and the dogs induced to follow them. When the bubbles are touched by the dog with nose or paw, they burst. Now, in the great majority of cases the dog takes little or no notice of the bubbles, and when they burst shows very little sign of surprise, or any other emotion. The game is evidently regarded as a poor one. Sometimes, however, when the dog bursts the bubble with his nose, he shakes his head, and appears to experience discomfort. In only one case have I found anything like a marked effect. The dog retreated under the furniture, and could not be induced to approach the bubbles. If one bursts one of the bubbles with one's own nose, one finds the result a little curious,—one feels the bursting, and there is a sudden concentrated flooding of the nose with a smell of soap. Such a gush of odoriferous particles may well disquiet an animal with so strong a sense of smell as the dog. I am indeed surprised that the majority of dogs take so little notice of this effect, though they sometimes do, as I have said, show signs of discomfort. I am disposed to think, however, that in the case of the dog which hid under the sofa there may have been association; for the dog hated being washed above all things, and would have to be dragged out of hiding when the time for the trying ordeal

* *Nature*, vol. xvii., p. 168.

came. Of all smells few had for him more unpleasant associations than that of soap. This of course may have been a coincidence. And I have tried other dogs which hated being washed without observing any such marked effect. One Yorkshire terrier, I may mention, licked the bubbles, and apparently liked this new method of experimentation. Its mistress told me that the dog liked soap, and always licked itself all over after the bath. Mr Romanes's fetishistic interpretation of the very marked effects on his terrier, I am disposed to question. He tells me that the dog disliked being washed, and thinks that such an association as I suggest was not unlikely.

It would not be difficult to fill several pages with examples of association in animals; but it is better to leave the reader to draw upon his own experience for supplementary cases. If he has any close acquaintanceship with animals, he will have little difficulty in doing so. Of course it is only when the idea suggested through association expresses itself in action that we can obtain evidence of its existence. And much of the training we give to our domestic animals consists in establishing certain associations, such that a word, a sign, a touch of rein or whip, suggests infallibly the appropriate action. And every one who has had anything to do with the training of animals knows well how much they differ in the rapidity with which association-links are formed, and in their permanence when once they have been established.

Further evidence of the establishment of associations is given where the animal in some way indicates to its fellows or to man its emotional state, or its desire that some action should be performed. I find that the sounds emitted by young chicks are decidedly instinctive,—that is to say, they are inherited modes of giving expression to certain emotional states. And some of them are fairly differentiated. At least six may be distinguished. First, the gentle piping sound expressive of contentment,—for example, when one takes the little bird in one's hand. A further low note, a sort of double sound, seems to be associated with extreme pleasure, as when one strokes the chick's back. Very characteristic and distinct is the danger note. This is heard on the second or third day. If a large humble-bee, or a black beetle, or a big lump of sugar, or in fact anything largish and strange, be thrown to them this danger note is at once heard. Then there is the piping sound, expressive apparently of wanting something. It generally ceases when one goes near them and throws some grain, or even only stands near them. My chicks were accustomed to my presence in the room, and generally were restless, and continuously made this sound when I left them. Then there is the sharp squeak when one seizes a chick against its inclination. Lastly, there is the shrill cry of distress, when, for example, one of them is separated from the rest. I have very little doubt that all of these sounds have a suggestive value of emotional import for the other chicks. Certainly the danger note at once places others on the alert, and the pleasure note will cause others to come to the spot where the little bird is when the note is sounded. I mention these associations, because they are the best examples I know of anything of the nature of inherited associations, and I would specially draw attention to the fact that they are at first scarcely at all particularised. The sound is associated with an emotional state, not with any particular experience which gives rise to that state, and it probably suggests a similar state not in any way particularised. It is possible that with the advent of particular experiences the sounds may be further differentiated; but I have not been able to obtain evidence of the fact. Somewhat will be said later on concerning the means of communication among animals; and the question how far such communication is suggestive of general emotional states, and how far of definite impressions or particular experiences, will be briefly considered.

Passing to higher animals, there is no question that they may be taught to signify to us by certain actions that they desire certain things to be done for them; and as this is evidence of particular associations, one or two instances may be adduced. One of my own cats would always touch the handle of the door when she wished it opened. We had established this association by building on an apparently chance occurrence, when the cat, reaching up, touched the handle. By always opening the door when she so reached up to the handle, and taking no notice when she merely sat and mewed, the suggestive value of the action became definitely fixed. Sir John Lubbock taught his intelligent black poodle Van to distinguish between plain and printed cards. The cards were about ten inches by three, and had some simple word such as FOOD or OUT printed on them in large letters. Van learnt not only to distinguish these from each other, but to associate particular words with particular experiences. When she wanted food or tea, she fetched the appropriate card and got what she wanted in exchange. In 113 cases, during twelve days, she only made two mistakes, one of which was bringing "door" instead of "food." In these cases she had twelve cards from which to select. These interesting experiments show differential associations involving considerable power of discrimination in sense-experience. To these cases I may add an observation which I have already given in my work on Animal Life and Intelligence. When I was at the Diocesan College near Capetown, a retriever, Scamp, used to come and sit with the lecturers at supper. He despised bread, but used to get an occasional bone, which he was not, however, allowed to eat in the hall. He took it to the door and stood there till it was opened for him. On one occasion he heard outside the excited barking of the other dogs. He ran round the hall, picked up a piece of bread which one of the boys had dropped, and stood with it in his mouth at the door. When it was opened, he dropped the bread and raced off into the darkness to join the other dogs.

Experiment and observation on our domesticated animals thus bring out abundant evidence of the association of ideas, and especially of new and unwonted associations. But it must be remembered that the whole natural life of animals affords a body of evidence which is only less striking because it is more familiar, and because we are so apt to suppose that the activities of animals are performed merely in accordance with their instinctive nature, forgetting to how large an extent this basis owes its differentiation to individual acquisition through experience. Amid more or less uniform surroundings, and under the influence of common habits of life, the individual experience is so far uniform as to give rise to similar associations. But this does not alter the fact that for the individual these particular associations are a matter of individual acquisition.

There can be little question that in animal life the vast majority of associations are associations by contiguity. But it is worth inquiring whether among animals, as with man, there are also associations by similarity. We may, at any rate for the present, exclude associations by perceived similarity of relationship. These depend upon the perception of relations, and I propose to devote a chapter to the question whether animals are able to perceive relations as such. The question is therefore whether ideas are suggested to animals through association by resemblance. When my chicks had learnt to avoid hive-bees, mainly I think through tasting or perhaps smelling the poison, I threw among them one or two of those flies, *Eristalis*, which bear a tolerably close superficial resemblance to the hive-bee. Not a chick would touch them, and the danger or warning note was frequently heard. Here was a clear case of avoidance through resemblance. But whether there was the suggestion of the idea of a hive-bee through resemblance is another matter, and one we are hardly in a position to determine. The question is whether the sequence in the consciousness of the chick was (1) visual impression of eristalis, (2) visual idea of bee, (3) taste-idea of poison; or (1) visual impression of eristalis, (2) taste-idea of poison. In other words, did the chick, on seeing the eristalis, have an idea of a hive-bee and then a suggestion of the nasty taste, or did the chick on seeing the eristalis have at once a suggestion of the taste? I think it far more probable that the chick simply mistook, as we say, the eristalis for a bee,—that, in fact, the impression of eristalis was nowise differentiated from previous impressions of hive-bees.

This case is typical of a whole range of biological phenomena which come under the head of mimicry. Mimicry involves resemblance; and in all cases the mimicking form (and it must be remembered that it is biological mimicry, not psychological: there is no possibility of intentional imitation) gains advantage from its resemblance to another form which is possessed of disagreeable or hurtful qualities. The resemblance also must be sufficient to suggest the hurtful quality as an idea of sense-experience. If the animal could differentiate the mimicking from the mimicked form, all the advantage would at once be lost. Hence it is probable that there is in these cases no association by resemblance, but rather a perfect illusion, the mimicking and mimicked form giving rise to a single undifferentiated impression. And we may fairly regard the glaring colours of noxious insects as specially developed, so as to be of strikingly suggestive value.

If then we conclude, as I think we must conclude, that cases of mimicry do not afford a trustworthy indication of association by resemblance, we shall find it difficult to obtain evidence of a satisfactory kind of the existence of this mode of association in animals. Not that this necessarily shows that such associations by resemblance are absent from the mental processes of animals. It is quite possible, nay more, exceedingly probable, that they may frequently occur. And the difficulty we have in finding evidence of their existence is perhaps due to the fact that such associations have little or no practical utility. With us association by resemblance has at most a descriptive value,—as, for example, when I describe Arthur's Seat, near Edinburgh, viewed from the Dalkeith Road, as resembling a lion *couchant*, and as not unlike the Lion's Head and Rump near Cape Town; or when the Neapolitans say that the steam-column of Vesuvius during an eruption is like a pine tree; or when I say that the taste of a pine-apple drop reminds me of the flavour of bananas. For animals this descriptive value is of necessity absent, and for them therefore associations by resemblance have little or no practical utility. It should of course be remembered that what I have just said of the merely descriptive value of association by resemblance applies to them only, and not to associations by perceived similarity of relationship, which have a rational value in the explanation of phenomena of far-reaching worth, and form a leading characteristic of insight. But these I believe have no place in the association of ideas as it occurs in animals.

CHAPTER VI.

MEMORY.

IN likening consciousness in its constant onward progress to a wave, we have noted that, besides the impression, or idea which occupies its summit or crest, it comprises also, at any moment, certain rising or waxing elements, and certain falling, fading, or waning elements. The waxing period is of variable length, and is difficult to estimate. When a stimulus is received on the nerve-ends of any sense-organ, there elapses a certain time before clear consciousness results. Some of this time is occupied in the transmission of the impulse along the afferent nerve. This is altogether outside consciousness,—a purely physiological matter. But when the conscious centre is thrown into activity, there is a shorter or longer space of time occupied by the development of the impression in consciousness, or, as we may presumably express it from the physiological point of view, by the rising of the cerebral disturbance into dominance. Impressions or ideas which are simple and familiar develop rapidly; when they are complex and unfamiliar, they develop more slowly. Even if we have a fair reading acquaintance with a foreign language, we find that it takes longer to read a page of that language than a page of our own. The somewhat unfamiliar words suggest ideas less readily than those of our native tongue. There is also a wide range of individual variation in the rate of development of ideas. Some people are naturally slow; others naturally quick. It does not necessarily follow that those in whom ideas develop slowly are inferior in mental worth to those in whom ideas develop rapidly. Sharp quick boys and girls do not always make original and exact thinkers. We may say then that the rate of development of the impression or idea,—remembering that these terms are applied to that which forms the crest of the psychical wave, or that which is in the focus of consciousness,—depends (1) upon the mental constitution of the individual, (2) upon the familiarity of the impression or idea, and (3) upon its complexity.

The fading or waning of impressions or ideas along the hinder slope of the psychical wave is a familiar fact, and one of great importance. Impressions or ideas do not cease instantly, but fade gradually. The meteor that shoots across the sky leaves a subjective track of light, due to the gradual fading of the earlier impressions. The continuity of all our impressions and ideas is of like order to the continuity of the meteor track. If the elements of consciousness were absolutely instantaneous in their occurrence and their cessation, the whole character of our mental life would be different from that of which we have practical experience. It has already been shown that the force of any word in a sentence is largely determined by what has preceded it. But if the occurrence and cessation of each idea were instantaneous, the preceding idea could in nowise influence the succeeding ideas. Or take the case of mental arithmetic. We could not follow out even a simple series of calculations if the result of one process were not carried forward to form the basis of the next. Suppose we say, for example: Take 12; add 15; divide by 3; answer 9. The idea of 27 as the result of the first addition must be still present to consciousness when the idea of dividing by 3 is suggested. It is this overlap which is one factor in giving continuity to such a process, and to consciousness and thought in general.

The rate of fading of different impressions and ideas is

probably very different. The fading is rapid for some sense-stimuli. Suppose that a series of similar sound-stimuli fall upon the ear. It is clear that they cannot be distinguished as separate unless the effect of the first has faded from the sense before the next comes. When the sense-fading (as we may term this fading of the results of simple stimuli) is *nil*, we cannot perceive any interval between them,—*i.e.*, they fuse into a continuous sound. But Von Helmholtz has shown that we can distinguish 132 musical "beats" in a second; and Professor Exner distinguished two successive snaps of an electric spark when the interval between them was only $\frac{1}{500}$th of a second. With the eye the sense-fading is very much slower. If sparks follow each other with an interval of less than $\frac{1}{25}$th of a second they appear to be continuous. In the case of touch impressions the results of observation are somewhat discrepant; but it is probable that when more rapid than from 25 to 35 in a second they are felt as continuous.

In these cases the sense-impression in consciousness is of the simplest possible nature, corresponding, we may suppose, to a brain-disturbance restricted in area and of little complexity. Exact measurements of the rate of fading of more complex impressions and ideas are at present impracticable. We are thrown back on introspection and retro spection, and on general considerations. When we read a paragraph or write one, or are speaking, we are more or less distinctly conscious as we read, write, or utter one word of much that has gone before. If a speaker to whom we are listening begins a sentence in one construction and ends it in another, the former has not faded from consciousness, and we are aware of the discrepancy. When the sentences are long, and overburdened with subsidiary clauses and parentheses, we not infrequently find that the beginning has faded before the end is reached. This is more markedly the case with sentences as heard than as read. In reading, a rapid glance of the eye back on the track it has been following serves to revive the fading ideas. But in listening to a speech or a lecture this is impossible. Hence in a lecture or speech construction quite admissible in the essay sounds laboured and awkward. Hence, too, in speaking to and writing for children and comparatively uneducated people the sentences should be short, and the construction direct and simple.

Ideas which are strongly touched with emotion are those which fade most slowly. A stupid or unintentional breach of good manners or social etiquette—a thing better left unsaid—will hover in the background of one's consciousness, and haunt one unmercifully for a whole evening. The news of a great sorrow abides in the mind, and tinges all our thought with sadness. The accepted lover sees everything through rose-tinted glasses.

We may say, then, that in this matter of fading, the impressions which result from simple and oft-experienced stimuli fade rapidly; that complex impressions and ideas fade more slowly; and that ideas which are strongly tinged with emotion linger longest in consciousness. And we may fairly suppose that the physiological conditions on which this psychical fading depends, are to be sought in the continued duration of the cerebral disturbances which accompany them.

It is clear that in any sequence of impressions and ideas the rates of fading of the several ideas will be different, or, to express the same fact in other words, their duration as marginal elements in consciousness will be different. In reading a paragraph, it is only the salient ideas that are carried on as marginal constituents of the psychical wave; the subsidiary ideas, and minor details, drop out in order of their lack of importance. And this differential fading, as we may term it, thus allows of the condensation of the net result of a long series of ideas in the final state of consciousness in which the series culminates.

The fading of impressions or ideas is sometimes spoken of as "primary memory." It is better, however, to restrict the word *memory* to the reinstatement or revival, through secondary suggestion, of psychical elements or constituents which have faded from consciousness. Thus the sight of violets has for me a tendency to call to mind, in memory, a particular spot in a quaint old garden, often visited in childhood; that is to say, violets have for me a tendency, through association of ideas, to form the focal starting-point of a purely representative visual scene.

Now in what respects does this purely representative visual scene differ from the presentative visual scene in which the violet before me is now set? In the first place, it lacks that vividness and insistency which characterizes direct presentation, and which is due to immediate excitation through nerve channels. But this is by no means all. Those who have exceptionally strong visualizing power can make a representative focal image more vivid and insistent than the presentative scene on which they are vacantly gazing. For most of us, however, our representative images lack vividness, lack detail, lack body, and, if I may so say, do not bear close and minute examination. One cannot focus first one object and then another in a scene representatively imaged,—or at least I cannot do so. Closely connected with this point is a second difference between representation and presentation, namely, that the former carries with it a certain foreignness to the subconscious and marginal presentations of the moment. I look up from my page, for example, and have a clear idea (image) of the bunch of bananas I saw on the table this morning at lunch. It carries with it a margin of the dessert dish on which the fruit lay, and other more or less dim surroundings. And all this is quite foreign to the subconscious presentations of the moment,—my manuscript, my study table, the dog on the hearth-rug, and so forth. This foreignness or incongruity supports the lack of vividness and insistency in differentiating the representative image in memory from the presentative impression. And the more vivid and particular the idea,—that is, the nearer it approaches in these respects an impression,—the more vivid are the revived and marginal surroundings of that image, and hence the greater the foreignness to the presentative margin. In dreams, and in certain abnormal waking states, the presentative margin of normal experience is absent or suppressed; and *then* the distinction between the representative and presentative tends to disappear, and the images assume the semblance of reality. A third point in which the representative differs from the presentative is its evanescence. Images, as they are normally experienced, lack for most of us that permanence which presentations owe to the continuity of the impulses received through the afferent nerves. The representative image, under normal conditions, only for a short time occupies the focus to the exclusion of the presentative visual impressions. Finally, there is what we may term the rational background, which enables us to differentiate the representative, however vivid, from the presentative. We know all the while, in the margin of our consciousness, that the image, however definite, is an image. We are here dealing, it must be remembered, with normal sane experience. In abnormal or insane experience the rational background may be absent, or the foreignness to presentative states may be unfelt, or the evanescence may give place to permanency, and representative images and ideas may come to dominate the mind with all the insistency of presentative realities.

It must, moreover, be particularly noticed that the reinstatement rarely or never reproduces a previous state of consciousness. It may *contain* a more or less faithful reproduction of a previous state of consciousness, but this reproduction only forms *part* of the new state of consciousness. And this is a further reason why a representation, by secondary suggestion, no matter how vivid and insistant, is seldom or never, under normal conditions, mistaken for a direct presentation by primary suggestion. It only forms part of the present state of consciousness. Let $\begin{matrix}\mathrm{A}\\ \mathrm{b\ c}\end{matrix}$ represent a state of consciousness of which A is the focal element, b c the margin or body of the wave. Let this be vividly remembered at a subsequent time. The wave will then be represented thus $\begin{matrix}A\\ b\ c\\ \mathrm{d\ e}\end{matrix}$, d e being the new margin or setting at that subsequent time. It is clear that these two states of consciousness $\begin{matrix}\mathrm{A}\\ \mathrm{b\ c}\end{matrix}$ and $\begin{matrix}A\\ b\ c\\ \mathrm{d\ e}\end{matrix}$ are different. And that the latter is not (though it does contain) a reproduction of the former.

In the case just considered, the focal element A carries with it in remembrance its original marginal setting a b. I remember, for example, with many details of marginal setting, the cobra that I saw on a granite slab surrounded with protea bushes on the basal slopes of Table Mountain. If I meet a man in the street, whose acquaintance I made some time ago at the house of a mutual friend, I am reminded of the circumstances under which we first met, of the place of our meeting, and the time. His face suggests a particular scene or particular scenes in my past life. The faces of our familiar friends of long standing, however, do not as a rule suggest particular associations. They have been seen so often, and under so many circumstances, that the particular associations have become evanescent, and given place to divergent associations. So too with the familiar objects and occurrences of our daily life; they have ceased to have particular associations. The formula for revival, thus becomes $\frac{A}{d\ e}$, there being no particular setting of A reinstated; and this, not because A was ever experienced without a setting, but because it has been experienced in so many settings that there is no revival of any particular one of them.

It is clear that there are two prerequisites of reinstatement, —first, the previous occurrence of the impression, idea, or marginal element to be reinstated; and secondly, the retention of something whereby the subsequent reinstatement is rendered possible. The former requires no comment here. With regard to the latter, the retention, we have already seen that we cannot in strictness say that ideas or elements of consciousness are retained in and by the mind. We must fall back on brain-physiology for an explanation of the so-called retention of impressions or ideas. The ideas as such have ceased to exist; but the brain-structure has been modified in such a way that under appropriate conditions similar ideas will be again produced. An analogy will make this clearer. When we speak into a phonograph the tones of our voice are not hidden away in, and retained by, the cylinder of the instrument; but the wax or other material is indented, as a result of the incidence of the sound-waves, in such a way that it is capable of reproducing similar sound-waves at a subsequent time. So, too, the brain tissue is so modified by the nervous disturbances which are the accompaniments of an impression that, under appropriate neural conditions, they tend to reproduce similar nervous disturbances which are accompanied in consciousness by a reinstatement of the impressions in the form of an idea. It is in this sense only that we may speak of the retention of ideas.

According to some students of physiological psychology there are separate parts or cells, the so-called "memory-cells," whose special function it is to perform the office of retention. When an impression A is produced in correlation with certain molecular disturbances, an effect *a* is produced elsewhere in the brain. Reinstatement is on this view a reproduction of this collateral effect *a*. This hypothesis is, in my opinion, unsatisfactory, and without warranty in observed fact. It introduces an unproved and unnecessary complication in the nervous mechanism. It is more probable that the occurrence of the disturbance associated with A leaves a tendency to the indirect or secondary reproduction of a similar disturbance *A*, and that the impression and its reinstatement involve one and the same area of disturbance. Be this as it may, how the tendency is established, and how it is retained, is a problem with which we must leave physiology to deal, being content, psychologically, to accept the fact of such establishment and retention. Psychology does, however, tell us somewhat concerning retention. We find, for example, that individuals vary a good deal in their retentiveness. Macaulay had only to read over a list of wranglers, and he could reproduce the list long afterwards without effort or mistake. Few of us could do this: some would remember, say, half-a-dozen of the names which perhaps struck us as familiar or peculiar; others would remember one or two; many would remember none of the names. Retentiveness is, in fact, to a large extent a psycho-physiological datum; something given in the brain-structure and mental character of each individual; something which we can no more alter than we can alter the size of our heads, or to take what is *perhaps* a closer analogy the size of our muscles. By careful use and training we may develop our muscles within the limits assigned to them by nature. So, too, by careful exercise we may perhaps develop our retentiveness within the limits assigned to it by nature. Opinions differ, however, as to how far mere retentiveness (that is ability to retain apart from ability to recall) can be strengthened by use; and it is a matter that is exceedingly difficult to test.

In the same individual retentiveness varies somewhat with the state of the nervous system. Impressions evoked when we are fagged out and weary, are not so readily retained as those evoked when we are fresh and vigorous. The impressions and ideas of the morning and in the spring-time of life, are more readily retained than those of the evening and in life's autumn or winter. Old people often remember vividly the impressions and ideas of their youth; what they fail to retain is the recent impressions and ideas. Their brains reproduce what is already registered therein; but fail to effect new registrations.

Other things being equal, the more vivid the impression, the more interest it arouses; and the more we attend to it, the greater the probability of its being retained. An impression may be strengthened by keeping it for some time in the focus of consciousness which involves attending to it, by bringing it into focus through different sense channels, —as where we make a child look carefully at, pronounce audibly, and write down a name we wish him to remember; and by repetition. The impression is apt to fade in a shorter or longer time. Occasional revival of the impression or idea at progressively increasing intervals leads to its more and more perfect retention. These modes of imprinting ideas on the mind, or establishing particular modifications of brain-structure, are well recognised by teachers.

Ability to retain and ability to recall are generally regarded as two different psychological faculties; but they are closely connected. The recall or reinstatement is, indeed, the only evidence we have of retention. We can know nothing about ideas that are retained but are beyond recall. Close as the connection undoubtedly is, however, it is advisable to distinguish between retention and reinstatement.

First, we may notice the distinction between casual or desultory memory and systematic memory. Let us contrast the two. The possessor of a good desultory memory has no difficulty in learning to spell, easily commits to memory strings of dates or lists of grammatical examples or exceptions; remembers without effort the Latin, Greek, French, or German equivalents of our English words. Foreign languages are readily acquired, since the vocabulary thus presents little difficulty. Such an one remembers the addresses of all his acquaintances, and the birthdays of all his relations and friends. His mind is stored with a mass of miscellaneous information, which seems always within easy call, and ready for the purposes of illustration. He is generally a good talker, a delightful raconteur, a superb diner-out. If he be a man of wide reading, he has always an apt quotation ready to his tongue. He remembers the plots of all the novels he has read, the exact routes taken by travellers and explorers, the winners of the Derby since he was a boy, and even the changes of political opinion of our leading statesmen.

Contrasted with such an one, is the man who remembers none of these things. He enjoys a novel, but forgets the plot in a week. He has read his Shakespeare with delight, but could scarcely quote you a single line. He fancies his friend has left London, but whether for Newcastle, Ramsgate, or Penzance, he cannot say. He knows that he heard that piece of music exquisitely played somewhere, and at some time, and was told who composed it, but whether it is by Offenbach or Wagner he forgets. He has yearly to be reminded of his wife's birthday. But get him to speak of philology, or botany, or the history of philosophic thought, as the case may be, and he will astonish you with the vast stores of his learning, and the methodical sequence of his ideas. He forgets no fact which bears upon his special subject, and which can find a definite place in its system of ideas. Such is an extreme case of systematic memory.

There can be no doubt that a good desultory memory gives a man great advantages over his fellows. It is generally associated with wide and general interests; and is due partly to strong natural retentiveness, and partly to the rapid and firm establishment of association by contiguity, and by similarity (resemblance), in so far as such similarity is tolerably obvious and near the surface of things. For there is a tendency in minds of this type to be superficial rather than deep, picturesque rather than logical, descriptive rather than analytical, reproductive rather than original. The analytical thinker, as he listens to the genial conversation of such men, admires perhaps their rapid flow of ideas, their fund of anecdote, their descriptive power. But he is apt to ask himself what after all there is in all this that is worth remembering.

When a man's memory is chiefly of the systematic type, matters are remembered in and through their significance, their value as evidence, their bearing upon some hypothesis, in virtue, that is to say, of their perceived relationships. Associations by contiguity, or by superficial and casual resemblance, are neither strong nor enduring. It is those deeper associations of a rational or logical character, which we have termed associations by perceived similarity of relations, which alone have permanence. The two kinds of memory are not, however, by any means of necessity mutually exclusive. The wide generality of the one, and the narrow specialism of the other, are often combined, and must be so combined to form a really great and commanding intellect.

It is questionable, as we have already hinted, whether desultory memory, dependent upon association by contiguity and by superficial resemblance, is susceptible to any appreciable extent of improvement by training and exercise. But systematic memory which is, in the main, dependent upon logical or rational association through the perception of relationships is more open to improvement in this way. Even here, however, it is rather the establishment of a definite system and skill in its use that is capable of improvement, than what we may term native power of remembering. It is power of recollecting rather than power of remembering which we can cultivate. To remember is an involuntary process beyond control; while to recollect is a voluntary process requiring some effort. Systematic arrangement and logical association of ideas help us to recollect. And when habit has rendered recollection along certain systematic lines comparatively easy, such recollection becomes closely akin to natural remembrance, just as in other cases to be hereafter considered, the results of habit approach closely to the results of instinct. From this point of view desultory memory is for the most part an inborn faculty; systematic memory is to some extent an acquired habit.

In clear remembrance and recollection there are often definite associations of place and time. Place associations we need not here consider. But since memory deals with the reinstatement of past impressions as present ideas, it will be well briefly to discuss the question of time-localization. The cobra incident that I spoke of a few pages back has for me not only place associations, but time associations. It occurred during the latter part of my five years residence at the Cape. By an effort of recollection I can fix the time more exactly as the spring of 1882. Applying the term "time-localization" to the process of assigning to an occurrence its position in the remembered duration of my life, let us proceed to consider how such localization in time is effected.

The first point to notice is, that only in the present moment of consciousness are events spread out before us in what may be termed their real time-size or duration. When, in the moment of recall, past events are represented in consciousness they are foreshortened or telescoped. When I run over in retrospect the events of my last summer holidays, or of my visit to the Auvergne, occurrences separated by days or weeks pass before my mind's eye in a few minutes. If this were not so it would take us an hour to recall the events of any past hour of our lives. The foreshortening or telescoping is effected through missing out or forgetting all the minor details, and recalling only the salient points. And this faculty of forgetting details it is which makes retrospection as we know it possible. The telescoping is more and more marked the further we look back into the past. The events of this morning body forth in memory with considerable fulness of detail. The events of this day last week have already been merged in the blended perspective of the past. Of last year only certain salient points stand forth; and further back still the rapid glance of memory only discloses some of the more important milestones which mark the progress of life's journey.

Localization in time—as, for example, when we endeavour to recollect when such and such an occurrence took place—is the determining of the position of the occurrence with reference to the salient landmarks of our past experience. I received a telegram yesterday, and I wish to recall when it reached me. It was after my morning lecture, and before I had my lunch. That sufficiently localizes it. When did I first meet So-and-so? It was after I returned from the Cape and before I was appointed lecturer in Bristol. By focussing attention on this period I can localize more exactly. I had heard of the Bristol post, but had not yet been appointed.

Such, I take it, is the manner in which we habitually proceed when we wish to recollect when an event took place, or to localize it in time. It involves the taking in at a glance of the chief landmarks of our life's history; the reference of the occurrence to a position between two of these landmarks; the focussing of the attention on this intervening period and the further defining of the time-position of the event with reference to the minor landmarks thus brought to mind. But how are the main landmarks themselves localized in time? By reference to the other landmarks leading from our earliest remembrances and recollections to the immediate past and so to the present moment of consciousness. In this present moment of consciousness our primary experience of sequence and duration is gained. Our past life, which we can review in memory, is an extension backwards through retrospective thought of experience gained in the present moment of consciousness. Our anticipations of the future are a similar extension forwards of this experience. Anticipation is prospective representation.

I have so far said nothing about the aid afforded by dates in assisting us to remember when an event took place. Dates are indeed secondary aids—most convenient and helpful, but none the less secondary—to time-localization. They are also common reference-points, enabling us to compare our own landmarks with those of other people, and thus to form a time-scale that is not only individual but social. My reader knows nothing of the landmarks of my life; I know nothing of the landmarks of his life. But if I say that I returned from the Cape in 1883, he can at once localize the occurrence in time by reference to some event in his own life with which this date is associated. Thus the date is a common reference-point by means of which I may compare my own time-localization with those of any one else. It is social and not merely individual.

Of some of the salient events of my life's story I know the dates; of my going to school, my leaving school, my visit to America, my engagement, my marriage, and so forth. For other dates I have to calculate from these fixed points. Had I a tenacious desultory memory no doubt I should remember a larger number of such dates than I do. But I think that the generality of people are pretty much in my position in this matter.

So far I have been considering the past of which I have direct and individual experience. But far back beyond my own individual remembrances and recollection stretches the greater past of history with all its social importance and value. Of this I have no direct first-hand experience; I have only the indirect second-hand experience gained from books and oral teaching. Into this greater past, so as to cover symbolically its longer periods, I extend that conception of time which has been found serviceable in dealing with the short space of my individual life. Just as I can review, foreshortened in memory, the historical events that have taken place in my own lifetime, so I can review the events that took place in the eighteenth century, the times of Elizabeth, or the reign of William the Conqueror. And just as dates are convenient, and socially indispensable, for marking the salient points of the history of the last twenty years, so are dates of equal value for marking the salient points of the centuries gone by.

It must be remembered, however, that dates are symbolic. Symbolic of what? Of localization in time. It is only as an aid to correct time-localization that dates are of any value. The mere repetition of a date tacked on to an event through association by contiguity is almost valueless educationally if it convey no conception of the relation of the event to other events in time-sequence. I ask a boy when Jonson's "Every Man in his Humour" was published. He answers glibly 1596. I say: Was Shakespeare still living? He looks confused; and I continue: Was Cromwell dead? to his still greater confusion. And yet he can give me pat the dates of Shakespeare's death and of Cromwell's protectorate. He has been badly taught—parrot fashion. He has the materials for time-localization, but the conception has not dawned upon him, or been explained to him. Another boy, of whom the same question concerning Jonson's comedy is asked, replies: "I do not know the exact date. But it was after the publication of the 'Faerie Queen,' and before Bacon's 'Essays' appeared. It must have been between 1590 and 1597." That boy intelligently localizes in time.

From simple reinstatement through association by contiguity, to the accurate localization in time of an event which took place long ago, is a far cry. This does but illustrate the fact, however, that revival or reinstatement is a necessary condition of a wide range of mental operations. And here it will be well to draw a distinction, one which we shall find to be of the utmost value and importance, between memory as merely reinstating, and memory in its reflective and retrospective aspect as affording the material of thought and knowledge. The difference between the two is this, that memory as merely reinstating does no more than introduce into the psychical wave revivals of former constituents of the wave of consciousness; while memory as reflective and retrospective carries with it some reference to the *relations* of that which is revived—some reference to the how, the where, and the when. The sight of a man's face in the street may vividly recall through reinstatement a scene, perhaps at a dinner-table, in which he formed a focal object, and perchance some witty remark of his. But if I then recollect where I met him, when, and under what circumstances, there is something more than reinstatement. I look back, reflectively, and view the occurrence in its relations to other events. The introduction of the relations is the introduction of a new factor. Of these relations that which involves the conception of time has been illustrated. Localization in time is, as we have seen, viewing the event in its true time relations. It involves reflection and retrospection, or looking back on the past course of the psychical wave and assigning to the event its true position in relation to other events.

Or take the cobra incident that I have more than once

mentioned. I never see a cobra, or the picture of one, nor do I ever hear the word, without that scene on the slopes of Table Mountain flashing across my mind. But the re-instatement of the scene is one thing, and its localization is another thing. I cannot localize it, either in time or space, without a *reflective* glance on my past life, my sojourn at the Cape, and when I was there. Now, whenever we speak of clear definite and accurate remembrance or recollection of an occurrence, we mean not only reinstatement but localization in our life's experience, and hence we are apt to suppose that such localization is one of the invariable incidents of distinct memory. I am, therefore, the more anxious that it should be clearly grasped that this localization is not a necessary feature—not a feature at all—of the reinstatement as such, but is something added thereto.

We shall have in future chapters to draw with much care the distinction between consentience or sense-experience and reflective consciousness. In the former, in sense experience, there is memory as mere reinstatement without retrospective localization. Only in the latter, in reflective consciousness, is that which is reinstated viewed in its relationship to other occurrences and to the past course of the wave of consciousness.

CHAPTER VII.

MEMORY IN ANIMALS.

A VERY short chapter will suffice for the consideration of memory as we infer it to occur in animals. In the first place we may notice that the existence of memory is implied in the association of ideas: or rather in the occurrence of ideas at all. For ideas are representations due, in the main, to secondary suggestion. But a representation is a reinstatement as an idea of that which has been presented as an impression; and such reinstatement constitutes, as we have seen, one aspect of the phenomena of memory. If, therefore, animals have ideas at all—and if they have not we need not attempt to carry any further our investigations into zoological psychology—they must have memory, and there must be in them, as in us, some anatomical and physiological basis for what is popularly termed the retention of ideas.

Observation, moreover, shows that some animals have remarkably tenacious memories. The chick, which having once tasted the poison of a hive-bee avoided again taking a bee into its bill, showed memory not inferior to that of the child which having once been burnt avoids for the future contact with fire. When I had taken some of my chicks to the yard whence I had obtained the eggs, and had made such arrangements as seemed feasible for their being absorbed into a brood of about the same age and standing, I returned after two days to see how they fared under these new conditions of life. One of them was some little distance from the hen, and when I bent down with outstretched hand the little thing ran to me and leaped upon my palm, scratching at it in the expectation of grain. I do not say that it remembered *me*, but it certainly remembered a particular action of mine to which it had grown accustomed.

When I was at the Cape I used to take my two dogs up the Devil's Peak, an outlying point of Table Mountain. There were several places at which it was necessary that I should lift them from ledge to ledge since they could not scramble up by themselves. After the first ascent they always remembered these places and waited patiently to be lifted up. On one of our first ascents one of them put up a young coney and they both gave chase. Subsequently, they always hurried on to this spot, and though they never saw another coney there, reiterated disappointment did not efface the memory of that first chase, or so it seemed. I think the last time I took them up must have been about three and a half years after the coney-hunt: so long had the memory endured and the association remained uneffaced.

Every one probably remembers the instance recorded by Darwin, whose dog recognised him on his return from his five years' voyage round the world in the "Beagle." Sir George Davis (1650) is said to have been recognised by a lion which he had brought up as a whelp, and parted with three years before. And Bingley* records another case when a lion showed indubitable signs of remembering a man who had been his keeper more than seven years before. These cases may be apocryphal, though I do not know that there are valid grounds for doubting them. In any case instances of memory in animals might be multiplied almost indefinitely; and any reader who has had opportunities of observing animals will be able, without difficulty, to supply additional instances from his own observation.

Those who have had some experience in training domes-

* "Animal Biography," vol. i., p. 240.

ticated animals will also be well aware that animals show a good deal of variation both in the rapidity with which they form new associations—that is to say, the frequency of repetition necessary to establish an association—and in the retention of the association when formed. Some dogs seem to learn rapidly, but soon forget the tricks one has taught them; others take longer to learn and seem to require that the association should be many times repeated before it begins to show signs of establishment: but when once the trick has been learnt they never forget it. I was informed by one who had some experience in training dogs for circus performances that those which were slow in acquisition were by far the more valuable for his purposes; and that very rarely did he find a dog that was both rapid in acquiring a trick and always reliable in its performance. But I am not sure that all trainers would concur in this opinion. All would, however, agree that dogs, even of the same breed and in the same litter, vary much in powers of memory.

The memory of animals must, I think, be entirely of the desultory type. Systematic memory involves, I conceive, powers beyond their reach. Whether this is so or not, however, depends upon the answer we give to the question to be discussed in a future chapter—Do animals perceive relations? I do not propose here to anticipate that discussion. It is sufficient to point out that systematic memory is based on the perception of relations. The associations formed are those by similarity of perceived relationship. A new fact is remembered not in virtue of its association with other facts or experiences by contiguity or by mere resemblance, but because it is seen to have a definite bearing upon other facts, to have recognisable connection with other experiences, to fit into a particular place in a definitely organized scheme, to have meaning in reference to a system of knowledge. Now knowledge, properly so called, is something very different from experience; and while association by contiguity suffices for and is the groundwork and foundation of sense-experience and the intelligence based thereon, association by similarity of perceived relationship is necessary to form the foundation of knowledge, and of all rational explanation of phenomena. Desultory memory suffices for sense-experience, and intelligence; systematic memory comes on the scene with the advent of knowledge and reason. Whether animals have knowledge depends upon whether they are able to perceive relations as such. If they have knowledge they may possess that type of memory which we call systematic—a type which arises out of and ministers to knowledge. But if they have, with all their wide range of practical sense-experience, no knowledge properly so called, then is their memory entirely of the desultory type.

Another question—whether the memory of animals may carry with it localization in time—again depends upon whether they can perceive time-relations as such. The discussion of time-localization given in the last chapter brought out the fact that this process—which we saw to be additional to, and no necessary factor in the process of reinstatement in memory—involves the location of the event, or experience in question with reference to, that is in relation to, other events in the time series. Is there any evidence that animals can thus localize events in time? Or let us put the question in another form: Can the phenomena of memory in animals, so far as we can observe and infer, be explained in terms of reinstatement, or must we infer a further process of time-localization? If they can be explained in terms of mere reinstatement, we are bound by the canon discussed at the close of the third chapter not to assume any higher process. For we saw that in no case is an animal activity to be interpreted as the outcome of the exercise of a higher psychical faculty, if it can be fairly interpreted as the outcome of the existence of one which stands lower in the psychological scale.

Let us take a typical case of memory in an animal. Captain Shipp gave an elephant a sandwich of cayenne pepper. He then waited for six weeks before again visiting the animal, and when he went into the stable began to fondle the elephant, as he had previously been accustomed to do. Watching his opportunity, the animal filled his trunk with water, and drenched the captain from head to foot. Now in such a case of definite memory of a particular occurrence, we are apt to suppose that the original experience is clearly localized, both in place and time. But a little consideration will show that this is not in the least necessary. All that the facts warrant us in concluding is that the sight of the captain gave rise, by association, to a reinstatement of a previous occurence. So long as the association had not faded from memory, it mattered not whether it was six weeks, six days, or six months ago; and there is no reason for supposing that the elephant in such a case localized the event in time, in even the most rudimentary way.

Animals seem however to show a sense of the lapse of time. A gentleman with whom I stayed at the Tijuca, near Rio de Janeiro, had a monkey (*Cebus* sp.), which was fed regularly at a certain time every evening. The animal was aware to within a few minutes when the time for his meal had arrived, and began to clamour by hammering on his board if the food was not brought punctually. This it may be said shows a pretty definite localization of time, for the monkey must have remembered that his meal was always brought at such a time. But surely the facts can be quite satisfactorily explained by the association of certain regularly recurring internal states with the regular satisfaction of hunger. And my friend told me that when the monkey was sick, he did not clamour for food; but that when he was recovering, he clamoured before the usual time. I think that many cases of the apparent sense of the lapse of time may be explained by direct association, especially when we remember how keenly observant animals are.

It is open to question however whether all such cases can be readily explained in this way. Take the instance, given in Mr Romanes' collection, of geese which used regularly once a fortnight to come into a little market-town on the morning after market-day, and to feast, "cackling and gobbling in merry mood," on the corn spilled by the millers from their sample bags. They always came on the proper morning fortnightly. "It happened," says Mr Goring, who reported the case to the *Gardener's Chronicle*,* "one year that a day of national humiliation was kept, and the day appointed was that on which our market should have been held. The market was postponed, and the geese for once were baffled. There was no corn to tickle their olfactory organs from afar, no traffic to appeal to their sense of hearing. I think our little town was as still as it usually is on Sundays. The geese should have stopped away, but they knew their day, and came as usual."

Such cases are difficult to explain on the association hypothesis, and whenever they occur should be made the basis, so far as possible, of experimental observation. It may be that there is true time-localization here, but I think it exceedingly doubtful. Those who have watched the mental development of children are well aware how very vague their time-localization seems to be. The past appears to remain for them undifferentiated, even into the nearer and the further past. And when we remember that it is only for purposes of description or explanation to our neighbours that time localization assumes any practical value and importance, we shall be more ready to accept the conclusion that for animals and very young children memory as merely reinstating is

* "Animal Intelligence," p. 315, from the *Gardener's Chronicle* for Aug. 3, 1878.

amply sufficient for all their practical needs. The exact conditions of reinstatement we may, in the case of certain animal performances, not be able to formulate. Nor is this to be wondered at when we consider the difficulties which stand in the way of the student of zoological psychology. One is forced back again and again, in one's endeavours towards accurate interpretation, on general considerations. And this question of time-localization is one of the many matters in the discussion of which we must endeavour to realize the conditions under which animals live. Depend upon it, their psychology is determined by that which is of service to them under these conditions. And it is difficult to see of what practical service time-localization would be to them, living as they do, so essentially in and for the present enjoyment of the passing moment.

But it may be said that since animals certainly localize in space, why should they not also localize in time? If it can scarcely be questioned that their memories carry with them the *where* of the original experience, why should we suppose that they fail to carry with them the *when?* Much however depends upon what is meant by their localizing in space, and what is meant by the memories carrying with them the *where* of the original experience. For there is a localization in space, loosely so-called, which involves merely association by contiguity in practical sense-experience; and a localization in space, properly so-called, which involves a perception of space relations as such. I fear that until we have discussed the nature of the perception of relations this may sound rather like the drawing of distinctions in which there lies no difference. There can be little question, I think, that when the elephant remembered the experience with the cayenne pepper, he remembered the circumstances. The reinstatement in memory presumably involved a representation of the whole visual scene. It involved quite possibly the reinstatement also of a sequence of events, foreshortened in memory—the entry of the captain into the stable, the giving of the sandwich, and the perhaps unusually hasty exit. All this would be part of the reinstatement through association by contiguity. But it differs from true localization, either in place or time, though it contains the raw material of both. For there is nothing but the representation of sense experience; there is no true perception of the relation of either category.

And here it will be well to leave these questions, until such time as we have occasion to reconsider them in the light of our discussion of the perception of relations. I will only ask the reader in the meanwhile to be good enough to credit me with an unbiassed desire to interpret the phenomena of animal psychology without exaggeration, either in the direction of excess or defect of mental power and differentiation. As an evolutionist who believes that the whole range of the mental faculties have been developed by natural processes, the tendency of my bias would assuredly not be in the direction of setting a gulf between the faculties of animals and the faculties of man. My sole aim is to endeavour to reach by legitimate process of scientific induction the most probable interpretation of zoological psychology, and by comparing this with the psychology of man to ascertain by what steps the lower faculties of animals may have passed by natural process of development into the higher faculties of man.

CHAPTER VIII.

THE ANALYSIS OF IMPRESSIONS.

WE have already seen that an impression is that which is brought to the focus of consciousness as the direct result of the excitation of afferent nerves; and that an idea is that which is brought to the focus indirectly, that is to say, through the intervention of an impression or of another idea. In the case of the impression, the concomitant cerebral conditions are due to afferent impulses coming from outside the brain; in the case of the idea, the concomitant cerebral conditions are determined from within by preceding brain-states. Impressions are therefore conveniently spoken of as *presentative*, as contrasted with those *representative* states we call ideas. Neither presentative impressions nor representative ideas constitute by themselves, however, states of consciousness. They are the focal constituents, it is true, but they are set in a margin of subconscious elements, which not only form integral parts of the states of consciousness, but also serve to modify, and in part determine, the character of the impressions and ideas themselves. And these marginal constituents are also in part presentative and in part representative.

We must now direct our attention more closely to the impressions of which the ideas are representative, and learn what we can, through the application of analysis, concerning their nature and origin.

Let us begin by inquiring what is presented to consciousness when we receive through the channels of sight an impression of such an object as a violet. It is necessary here to distinguish between that which is directly presented to consciousness and that which rapidly follows on this presentation. We actually see the violet as coloured, of a certain figure, and definitely located in space; all this seems to be part of the direct impression. Its scent, its weight, how it feels in the hand, the taste of a petal in the mouth, its name,—these or some of them, and perhaps more besides, may be rapidly suggested by the sight of the flower; but they seem to be representative and added to the direct impression. They are suggested through association, and are the results of experience. For the child who has made acquaintance only with wild dog-violets, the sight of the flower will not give rise to any representative scent element, such scent not having been associated with the flower through experience. It would seem, then, that the presentative impression, as such, does not contain more than elements of colour, of form, and of position in space. May it not, however, contain less? Is all this indubitably presentative? Is not, for example, the position in space something superadded to the direct impression? For the adult consciousness it would appear not. We cannot get a definite visual impression of an object without this element. We cannot see a violet without seeing it somewhere. We do not first see it and then locate it in space. When we see it at all we see it out there. It would seem indeed that all our visual impressions carry with them this externality. Even when we lie on our backs and gaze up into the cloudless blue of a summer sky, we feel that there is in the impression an inalienable element of depth.

Externality is also given us in auditory impressions, though our ability to localize the source of such impressions is here very markedly less than in the case of sight. When we hear a sound there appears to be presented, with the tone or noise, a certain outness; and this does not seem to be merely suggested by the impression, but to be part and parcel of the impression itself. In olfactory impressions and impressions of radiant heat, on the other hand, the element of outness seems to be very vague, and to be rather suggested as the result of oft-repeated association than directly presented and involved in the impression as such. In tactile impressions there would seem to be an immediate reference outwards to the part of skin-surface affected, while in the case of gustatory impressions a general localization in the neighbourhood of the mouth is probably equally immediate.

Of all impressions of the special senses, visual impressions are the most clearly defined, and for human beings are the dominant impressions, at least so far as the information they convey is concerned. Touch is sight's constant ministrant, ever ready to verify the visual impressions. Hearing widens the field of consciousness, and for us, through oral communication, is a channel of universal suggestiveness. Smell, taste, and the temperature sense are in this connection of secondary importance. Let us then return to the visual impression and endeavour to analyse it further. Inseparable though the form and distance of the violet may be from its shading and colour, we nevertheless feel that these several elements are distinguishable on careful introspection. Nay, further, we shall learn that, inseparably combined as they are in the psychological impression, analysis decomposes them into elements which are *physiologically* separate and distinct. In this inquiry, indeed, it is necessary to dig down to the organic and physiological foundations in order that we may adequately grasp the nature of the psychological superstructure. We must even go further and see how these physiological foundations are determined by physical conditions external to the organism.

I said just now that on careful introspection the shading and colour of the violet are distinguishable from the elements of form and position in space. Of these, however, shading and colour seem much more intimately united together than is either of them with distance or position in space, while form appears in some respects to be intermediate in character. In so far as the form involves only extension in two dimensions, on a plane at right angles to the line of vision, it appears to be closely associated with the shading and colour; but in so far as it involves solidity, which introduces the third dimension of space, it appears to be more closely associated with position in space. Leaving this solidity to be considered presently, together with distance and position, we will now direct our attention to the extension, shading, and colour of the violet or other such visible object in the external world. Physical science tells us that the visibility of the object is dependent on the surface reflection of rays of light, which are transmitted in right lines to the corneal window of the eye through the medium of the ether. A beam of sunlight consists of transverse vibrations of the etherial medium of almost inconceivable rapidity, of which all that lie between about 800 billions per second and 400 billions per second are capable of affecting the retina of the eye. There is a long series of vibrations of less rapidity than 400 billions per second which may be felt as radiant heat; and there is a long series of vibrations of greater rapidity than 800 billions per second, of which we have indirect knowledge through their chemical effects on certain substances; but neither series is capable of affecting the retina. By means of a spectroscope the vibrations which do affect the eye may be arranged in a continuous series, being spread out in the order of their frequency from 400 billions per second at one end to 800 billions per second at the other end, so that between the extremes there is a physical gradation through a great number of minute stages. When a beam of sunlight comprising vibrations of all stages of frequency between and beyond these limits, falls on such an object as a violet, the vibrations are partly absorbed and partly reflected. In the petal of the violet there is relative absorption of the waves of less frequency, and reflection of the waves of greater frequency. But all that reaches the retina from the violet, or other external object, is a more or less selected sample of etherial vibrations lying between the limits of visibility. Such, in briefest outline, is the physical story of that which goes on external to the organism. Let us now turn to the physiological story of what takes place when the etherial vibrations reach the organism.

When the vibrations, transmitted in right lines, reach the eye, they pass through the convex transparent cornea, through the pupil surrounded by its iris curtain, through the crystalline lens and vitreous humour, and are thus made to impinge on the retina. All these transparent structures are merely accessory, and for the purpose of throwing a clear-cut, well-defined vibration-image on the essential receiving structure, the retina, which is in direct communication, through the optic nerve, with the brain. The retina itself, in its focal and most sensitive area, is composed of an exquisitely delicate mosaic of minute cones, of which there are some 2,000 in the restricted area, less than $\frac{1}{100}$th of an inch in diameter, of most acute focal sensibility. Outside this area there are, in the retinal mosaic, delicate rods as well as cones. Each of these minute cones and rods has probably its individual connection, through the fibres of the optic nerve, with a particular part of the brain; and each is physiologically sensitive (using this term in a non-psychical sense, as a photographer speaks of a sensitive plate) to light vibrations. Since, therefore, the retina is an extended surface, on which is thrown a vibration-image having extension, we have here a physiological basis for that extension in two dimensions with which we are psychologically acquainted. Only that part of the retinal mosaic on which there falls the image of the violet, or other external object, is stimulated by the etherial vibrations. So far there is little difficulty; and if the external world appeared of a uniform grey of merely varying intensity, we could sufficiently account for the shading of the extended surface by differential stimulation of the retinal cones or rods. But objects appear not only shaded in neutral grey, but tinted of many hues. Herschel thought that the workers on the mosaics of the Vatican could distinguish at least 30,000 shades of colour. Now there is no reason to doubt that each retinal cone is capable, when stimulated, of initiating the impulse which gives rise to any of the numerous colour sensations. There are not some cones set aside to respond to certain particular vibrations, others set aside to respond to other vibrations, and so on; each cone may respond to any vibration within the visible scale. But it is clear that it does not respond in only one way to vibrations of all frequencies. If it did this the world would appear neutral tinted, without variations of colour. Nor, on the other hand, can we suppose that it contains some thousands of substances of differentiated sensibility, each of which is stimulated by vibrations of a definitely assigned frequency or range of frequency. It seems that the phenomena can be best explained if we assume that there are in each cone a few primary modes of response, by the subtle combination of which all the diverse colour-effects of experience may be produced. It is found, for example, that if we take vibrations of a certain frequency in the red end of the spectrum, others of a certain frequency in the green, and others of a certain frequency in the blue, we can, by suitable combinations and by varying the relative intensities of these three, match very fairly any colour or shade which we meet with in normal experience. If we similarly take four primaries, red, yellow, green, and blue, we can, it is said, make a somewhat better match. Into the physiological theories of colour-vision, however, we cannot enter. By the development of what organic products (or in what other way) these different modes of response are rendered possible; whether there be three or more fundamental modes of response in the retinal cone; whether the sensation of white is physiologically simple or compound; whether black is merely negation of response or a separate mode of response,—these are questions which must be left for physiology to answer. It must here suffice to note that they are questions under discussion, and ones to which a definite reply is by no means easy.

For psychology, vision begins with the visual sensations; and for us the primary question is, What is a visual sensation? A *sensation* may be defined as *an undecomposable psychological element directly due to an afferent impulse or combination of afferent impulses.* It is thus an ultimate result of psychological analysis as such. I would specially draw attention to the fact that the sensation is the result of analysis, and to the sense in which this statement holds good. The *data* of experience are states of consciousness. With them we commence our psychological study. We wish to know all that we can about them, and we therefore submit them to analysis. The first result of our analysis of a state of consciousness is to distinguish a focal impression or idea from a marginal setting of subconscious elements. We then proceed to analyse the focal impression, say of a violet, and we find that the impression is somewhat complex, involving colour, shading, form, and position. We therefore analyse further, and taking the element of colour we reach a sensation of violet, which appears to be psychologically undecomposable. Here is an end of our psychological analysis in this direction. But we may carry back the analysis further physiologically, and may show that the sensation of violet is due to the combination of physiological impulses, namely (to take one interpretation), to strong impulses due to physical vibrations of a certain rapid frequency, with feeble impulses due to vibrations of other frequencies, medium and relatively slow. The steps of the analysis, therefore, are—state of consciousness with which we start: impression: sensation: physiological impulse. The impression and sensation are psychological: the impulse physiological. Hence the sensation is the ultimate result of psychological analysis.

But now suppose that instead of starting from the data of experience, states of consciousness, we endeavour to explain how these states of consciousness have been built up, then we proceed to utilize the results of our analysis and make the elementary sensations our point of departure: and this method is so common in psychology, that the fact that the sensations *are* the results of our analysis of states of consciousness is apt to be lost sight of.

To return now to the visual sensations which psychological analysis discloses as elementary constituents of impressions. There can be no question that, for psychology, black and white are primary sensations. But what about grey? Can we analyse a uniform grey into black and white, as we can analyse a musical chord into its constituent tones? So, too, with such colours as orange or purple. When we look at a homogeneous patch of orange, uniformly illuminated, can we analyse it into red and yellow elements? or does it produce an effect which, as a sensation, is incapable of analysis however well we may *know* that it is the result of the combination of diverse physiological impulses? Opposite answers are given by different observers. For myself, I am quite incapable of analysing orange when it is presented to me as part of a definite impression. I happen to know that pink, to take another example, is the result of, so to speak, watering down carmine with white. But when I look at a uniform pink surface I am incapable of analysing the sensation. When I look at a purple surface I can see that the colour has certain affinities to blue and certain affinities to red. But I cannot pretend to analyse it as I can analyse a musical chord. For myself, therefore, and those whose experience is similar to mine, it would seem that grey, orange, purple, pink, and so forth, are primary sensations which result from the combination of physiological impulses below the threshold of consciousness. For us their "dissociation point," to borrow an analogy from chemical science, lies in the infra-conscious physiological region. This does not, however, preclude the view that for others differently constituted the combination occurs above the threshold of consciousness, and that they can analyse in a region into which the conditions of my conscious nature preclude my following them. For them pink is not a sensation, but the result of the combination of sensations of carmine and of white.

To sum up then concerning the physical, physiological, and psychological aspects of this matter, we have first in the physical field of external nature a long series of etherial vibrations capable of stimulating the retina with a wide range of graduated frequency; we have, secondly, in the physiological field of the retina, a few differentiated modes of response to groups of these etherial vibrations; and we have, lastly, in the psychological field of consciousness, a great number of different sensations, for some of us incapable of analysis in this field, due to varied combinations of the few physiological responses. These visual sensations are, however, merely elements in the visual impressions which carry with them other elements, such as solid figure and position in space, to which we shall have to return presently.

Auditory impressions, like visual impressions, carry with them, as we have seen, an element of outness or externality. Apart from this they consist of auditory sensations. Physical science tells us that sounds take their origin from some external vibrating body, and are conveyed to the ear as waves of elastic compression of the air. The vibrations thus transmitted are longitudinal or to and fro, not transverse as in the case of light. Within the external opening of the ear, these waves encounter the drum membrane which they set in vibration. Into the nature of the accessory parts, and of the essential structure of the auditory organ, we cannot here enter. Suffice it to say that by the accessory parts, which though organic are to a large extent mechanical in their function, the vibrations are transmitted to a fluid contained in the essential organ, which, itself bathed in fluid, lies in a cavity in one of the bones of the skull. In a certain part of this organ, termed the cochlea, there is a spiral membrane with delicate radial fibres, and this membrane is so supported and arranged as to be tense radially, but loose longitudinally, or in the direction of the spiral. The fibres of the spiral membrane are not themselves directly connected with the fibres of the auditory nerve, but they are connected indirectly through their supporting structures. If any particular radial fibres are caused to vibrate, their vibration stimulates certain hair-cells indirectly associated with them; and thus it is possible that every radial fibre sends its special and peculiar excitation along its associated auditory fibre to a particular part of the brain.

When we sing a note into an open piano, the loud pedal being held down, the sound waves caused by the voice, sweep over the strings, and tend to set them in vibration. But only those wires whose tones are in unison with the note we are singing are thrown into vigorous vibration. The other wires remain comparatively unaffected. And when our voice ceases, the note is continued through the vibration of those piano-wires which answer to it in period. If a *noise* be made near the piano, such as the striking of a book with a stick, a confused murmur will be heard among the strings. For the noise is a chaotic mixture of vibrations, and from this each string mechanically selects the vibrations whose period answers to its own. Now it seems probable, though there are still unexplained details, that the spiral membrane is in some manner selective like the wires of our recipient piano. When a note falls upon the ear, vibrations of the fluid are made to sweep along the membrane; that part of the membrane which is attuned to the vibrations is affected, and through it the associated hair-cells in connection with auditory fibres are stimulated. When a noise falls upon the ear a considerable area of the membrane is set in vibration, and a somewhat chaotic set of impulses travel together along the fibres of the auditory nerve.

The range of vibrations capable of stimulating the auditory membrane varies a good deal in different individuals. About thirty vibrations per second is the lower limit; the upper limit is about 20,000 vibrations per second, but some individuals have their range extended to 40,000. It should be clearly noted that although *physically* there is a close analogy between the series of light-vibrations of increasing frequency, and the series of sound-vibrations of increasing frequency, yet *physiologically* there is this great difference. Any light-vibrations belonging to the visible series can and may stimulate each retinal cone, and the differentiation for colour is somehow effected within the cone itself or in some structures connected therewith on the way to or within the brain. Sound-vibrations on the other hand affect, according to their frequency of vibration, different auditory hair-cells. Hence the combination of visual impulses to form a physiologically compound colour such as purple is of a different order from the combination of auditory impulses to form a chord. Psychologically I am unable to analyse purple; the combination of impulses is for me infra-conscious. But I can quite easily analyse a chord into its constituent notes; the dissociation point of a musical chord is within the conscious region. I must, therefore, speak of a combination of impulses (physiological) to form the sensation purple; but I can speak of a combination of auditory sensations (psychological) to produce a musical chord. Noises, however, from their chaotic nature are only to a slight extent, and with difficulty, analysable.

Closely associated organically with the auditory organ—nay forming part, and almost unquestionably the original part of the structure long known as the auditory sac or membranous labyrinth of the ear—lies the seat of certain very different sensations; namely, those by which we are made sensible of changes in the direction or in the amount of the movements of the body as a whole, and of the head. The exact nature of the physiological impulses which conspire to give rise to these sensations is not thoroughly understood. No doubt the impulses orginate in movements in the fluids which fill, and surround the membranous labyrinth with its semicircular canals; but exactly how these movements stimulate the end-organs of the afferent nerve-fibres is not known with certainty. Even the sensations themselves, though they are now recognisable without difficulty, are so little obvious in ordinary experience, that, like the motor sensations to be considered presently, they play little or no part in the older psychology. We may be said, indeed, to owe their recognition to the alliance between physiology and psychology, and to the introduction of methods of experimental investigation.

If a man sit blindfolded on a smoothly running turn-table he feels a curious sensation of being turned when the table is rotated; but if the rotation be continued at a uniform rate the sensation dies away, but is again felt whenever the rate of motion is accelerated. If, on the other hand, the rate of rotation be diminished he has a sensation of being turned round the other way. If he sit on a swing, blindfolded, and the swing be gently moved he is at once aware of the fact. If he be on a smoothly running car, he feels each change of rate of motion. And if it run for some time at a uniform rate and then slacken speed, he feels as if it were beginning to move in the opposite direction. Such are some of the experimental evidences of the real existence of these sensations. And I call them sensations because they do not appear to be psychologically analysable.

It will be unnecessary for us to analyse the impressions of smell, taste, touch, and temperature, or consider olfactory, gustatory, tactile, and temperature sensations. The foregoing analysis will suffice to show the psychological and physiological nature of sensations and impulses due to the stimulation of the end-organs of the so-called special senses. We may now note that in the impulses so arising there are differentiations of two kinds. There is a differentiation of similar stimuli according to the position of the stimulated end-organ on an extended surface, as in the case of touch. There is also a differentiation of dissimilar stimuli acting on the same end-organ as in the case of colour vision, probably of smell, and perhaps of taste. And these differentiations may be combined as in the case of vision, where each cone of the area of maximum sensibility is not only differentiated in position, but exhibits differential response according to the wave-frequency of the light by which it is stimulated; or again, as in the case of hearing, where the differentiation according to position is associated, through the intervention of the accessory structure of the spiral (basilar) membrane, with sound stimuli of a given wave-frequency. We have also distinguished between the afferent impulses, which are physiological in their nature, and the sensations to which they may give rise, and have seen that these sensations, and not the mere impulses, are the psychical elements which constitute those impressions that we experience. And now we proceed to note, as in accordance with all that we have already learnt, that at any moment there are raining in on the brain through the afferent nerves multitudinous impulses, some of which are definitely of conscious sensation-value, many of which are merely subconscious, and yet more of which are infra-conscious, merely physiological, and not entering the psychical wave at all. As I write there is, besides the focal visual impression, much that is merely subconscious in vision, and perhaps also visual impulses that are infra-conscious. But apart from these occurrences in the visual field there are very numerous tactile impulses. My clothes press gently on many parts of my body, my attitude involves stronger pressures, my fingers hold the pen. There are, too, numerous warmth-impulses from the fire; and many auditory impulses fall upon the ear, the ticking of my clock, the pleasant murmur and crackle of the burning log of wood, sounds of birds without, and of my pen running over the paper. There are also olfactory impulses from my cigarette and so forth. All these impulses pour in by the afferent nerves, and there arises from them a definite series of focal impressions and ideas suggested thereby, a more or less indefinite margin or setting to the same, and a physiological background which though infra-conscious is constantly so to speak hanging on the confines of the psychical wave.

So far I have said nothing of sensations of pain, though these are often unpleasantly insistent elements in consciousness. The word pain is here used in its narrower acceptation, not as antithetical to pleasure, but as descriptive of a particular kind of sensation. If the skin be removed from any part of the body, and the underlying dermis be exposed, a stimulation, however light, does not give rise to a sensation of touch, or to one of heat or cold, but to one of pain. If we increase the pressure on a small area of the skin beyond a certain amount we cause a sensation of pain; and if we either chill or heat a portion of the skin-surface beyond certain limits pain arises. It seems proved that the nerve-endings which are stimulated to pain are different from, and probably lie deeper than, those which are concerned in sensations of touch or of temperature; the nerves with which they are connected pursue a somewhat different course in the spinal cord, and end in different centres in the brain. In certain diseases, and under the influence of certain drugs, sensations of pain and sensations of touch and temperature are differently affected. It is worth noting that the sensations of pain, in the narrower sense, are not always painful or disagreeable in the wider sense. Moderate pain, as when we lightly press a bruise, or begin to make our hands tingle before the fire, is not by any means unpleasant.

These sensations of pain are developed not so much in the skin as beneath the skin; nor are they restricted to the neighbourhood of the skin. From tendons, bones, joints, and the internal viscera, we may receive sensations of pain, and the only definite impressions we receive from within our skins are those originating in sensations of pain. The healthy man, sound in every part, pays very little heed to what is going on inside him, apart from the calls of hunger and thirst and other natural promptings. There is probably nevertheless a very considerable body of impulses pouring in on the brain from all the inner parts. This is often summed up under the head of "general sensibility." It is this which, as we have before seen, plays no unimportant part in determining those marginal factors of our psychic life which we term mood and temperament, and which were indicated as part of the organic basis of personality. It seems very probable then that pain is felt when these sensations of general sensibility are abnormally heightened. Incipient heightening or the beginnings of pain are not unpleasant, but they soon reach a stage when they are more or less distressing. Fatigue is probably a modification of general sensibility, and in its incipient stages, when we are, as we say, just healthily tired, is not unpleasant. Hunger is associated especially with the condition of the mucous membrane of the stomach, and may range from a pleasantly healthy appetite to the gnawings of famine. So too with thirst, which appears to be due to a deficiency of water in the mucous membrane of the soft palate. Fatigue, hunger, and thirst, with the pain felt in the viscera, the eyes, ears, head, and so forth, and in or beneath the skin, are more or less diffused presentative states, which carry with them a varying degree of localization.

The agreeableness or disagreeableness of impressions or states of consciousness—that emotional colouring or tone which makes them pleasant or distasteful—should probably be regarded rather as psychical *qualities* of sensations than as separate and distinct elements. It will be sufficient to make but a passing reference to them here; their differential influence on the control centres will be briefly considered hereafter. The pleasurable tone sometimes tinges the sensation or impression; sometimes arises out of the relations of the focal and marginal elements of a state of consciousness; and sometimes is due to associated ideas, being of course in this case representative. A pure saturated colour, a full rich musical tone, a diffused sense of moderate warmth, certain scents and flavours; these are pleasurable in themselves. Other scents, other flavours, shrill high-pitched notes, pain in any of its intenser forms, are in themselves distasteful or even distressing. In the case of a musical chord or a sequence of such chords, in the case of a fine painting, the pleasure arises from the harmonious relations of the parts. And nothing is better calculated to enforce the complexity of the psychical wave in the instant of consciousness than a consideration of what elements are present, focal and marginal, when our soul is stirred by some magnificent pageant, by a rich and full orchestral climax, or by the culminating scene in a great play or novel. It may perhaps be said that in such cases we have the accumulated effect of states of consciousness which have succeeded one another during a considerable space of time. But if it be true that we are limited to the present instant of consciousness, it is clear that at such moments of climax there are gathered up into the wave of consciousness a great number of elements, the harmonious correlation of which gives us a pleasure of peculiar richness and fulness. Some of these are representative; and pleasurable or painful tone is often associated rather with the suggested ideas than with the suggestive impression itself. There is nothing intrinsically pleasurable in the cawing of rooks, nor in the shrill cry of the swifts as they wheel in their flight through the summer air; but these are sounds which for many of us are full of pleasant associations. The scent of violets in the early morning, the song of the lark in the summer sky, the sight of the young grass when the hay has been recently cut and carried, the warmth and brilliance of the sun after dark and cloudy days,—all these are pleasant; each has its special quality of emotional tone; and all are due to complex correlated influences, presentative and representative, but mainly representative. So, too, with disagreeable or distressing states. The painful horror of the sight of blood, the loathing some feel for a spider or a toad, the indefinable fear of a snake,—these are due to the emotional tone of associations which are often complex. Lastly, the exhilaration of perfect freshness and health, and the depression of weariness and dyspepsia, are due to the emotional tone in association with the net result of multitudinous impulses from the nerves of general sensibility.

There is yet one more group of sensation-elements that remains for consideration,—the group which constitutes what is often spoken of as the muscular sense. The name is not quite satisfactory; and it will be better to speak not of muscular but of motor sensations, including under this head sensations of position, such position being regarded as the resting stage of movement. We need not here discuss the physiological question as to what share of these motor-sensations is due to afferent impulses from the muscles themselves; what share is due to movements and strains in the skin or other surrounding parts; and what share is due to impulses coming from the articular surfaces of joints, or from parts which move over each other, as the eye moves in its socket. Psychological introspection would seem to assign to the last class of impulses—that of the movements of parts in their sockets—the leading position. But this does not exclude the other factors.

There can be little question that motor-sensations are capable of very exact definition, and that they are of great importance in building up the fabric of our presentative impressions, as they are also in affording data—to be considered hereafter—for the control of action in the field of practical skill. If with shut eyes and outstretched hand we trace a circle in the air, we shall perhaps be surprised at the accuracy with which we feel the movements and positions of the limb; or if under similar circumstances we draw a small circle on paper, or write our name slowly blindfolded, the delicacy of our motor feelings may, if our attention has not before been drawn to the matter, come upon us almost as a revelation. For it is characteristic of these motor-sensations, that they lurk somewhat persistently in the marginal region of consciousness, and require the application of our attention in an unwonted direction to be rendered focal. And this would seem to be due to the fact that the movement or position is in daily life merely a means to attain a given end, and that the end and not the means is what we habitually have in view. It is therefore worth while to carry out some self-observation directed to the ascertainment of the extent and limits of our motor-sensations. We are looking round us all day long with our eyes, and yet we may scarcely have realised how distinctly we can feel the movements of the eyes in their sockets. We are constantly adjusting our eyes for near and far vision, and yet may not have paid any attention to the motor-sensations which accompany these adjustments. Nevertheless, when our attention is drawn to the fact, we feel them distinctly.

We have seen that along the afferent nerves of the special senses, and again through the nerves of general sensibility, impulses are constantly pouring in upon the brain, some of which, presumably those which are dominant, enter into the focal impressions of consciousness; others, as subconscious elements, fill in the margin of consciousness; while yet others remain infra-conscious,—no doubt of great organic importance, but psychologically extra-marginal. So, too, there is pouring in upon the brain a multitude of motor-impulses, some of which, as we shall see in the next chapter, enter into and modify the focal impressions, many of which are at most marginal, while a residue, large and by no means unimportant, remains infra-conscious, and ministers to the physiological integrity of the bodily life.

It only remains to add that, while we speak of "a sensation" or "sensations" as undecomposable for consciousness, we often employ the word "Sensation" (without the article and not usable in the plural) as a grouping term for the whole of that area of mental life which is comprised under the head of sense-experience. Sensation in this usage is comparable in classification with Perception and Conception.

CHAPTER IX.

SYNTHESIS AND CORRELATION.

IN the foregoing chapter we submitted impressions to analysis, and we found that the ultimate psychological results of that analysis were sensations. We saw that these included sensations of the special senses, sensations of pain and general sensibility, and motor-sensations, any or all of which may carry with them an emotional tone pleasureable or the reverse. And, furthermore, we saw that the sensations might in physiological analysis be further decomposed into impulses,—the sensation purple, for example, being due to a combination of such physiological impulses. I have used such phrases as "the combination of physiological impulses to give rise to a sensation" to avoid pedantry and circumlocution. But it may be noted that such phrases are, in strictness, open to a very serious criticism. Physiological impulses are waves of molecular change of some sort, and are physical in their nature; while sensations are elements in consciousness, and psychical in their nature. No combination of physical impulses can give rise, it may be said, to a psychical or conscious condition, for there is no community of nature, and no thinkable continuity, between physical states and consciousness. With this criticism I am heartily in agreement. What I should say, in strictness, is, that the physiological conditions which are the concomitants of sensations are due to the combination of physiological impulses, the concomitants of which are infra-conscious, or do not rise above the threshold of consciousness. But such sentences as this are rather cumbrous! While, therefore, I heartily agree with the spirit of the criticism, I shall for the sake of brevity and clearness use the less accurate phraseology.

We have now to see how the sensation-elements are combined synthetically to form impressions as we know them; how they enter into correlation with each other; and how they call up through association representations of similar sensation-elements. It must be remembered that the web of consciousness, even in sense-experience, is wonderfully complex, with a complexity which beggars description. All that we can do in this chapter is to consider some of the more important of the correlations, and to indicate the manner in which sensations enter into synthetic union.

When we look at such an object as a cubical brick lying on a table at a distance of three or four feet from us, we obtain a definite impression. We are, however, using two eyes. On the retina of each eye there is an image of the brick; but the grouped impulses from the two retinas are factors in a single impression. Moreover, the two images are not quite alike, for the two eyes look out on the brick from slightly different positions, and a little experimenting will show that what is seen with the left eye is not quite the same aspect of the brick as that which is seen with the right eye. But not only do the different groups of impulses from the two eyes conspire to give rise to a single impression, they also give, or aid in giving, to the impression the specific quality of solidity or spatial depth. This may be experimentally shown with the aid of the stereoscope. By means of this instrument two different flat pictures, taken from slightly divergent points of view, are seen at the same time, the one with the left eye, the other with the right. The effect is a wonderfully perfect illusion that the objects have solidity, and that they stand out in space of three dimensions. Even if the pictures are only momentarily illuminated by the electric flash, the illusion of solidity is still conveyed. It would seem, then, that the synthetic combination of the impulses generated by two different retinal images, gives to the visual impression its element of solidity or depth. Now the question arises, have we in this synthesis a combination of physiological impulses, or a combination of sensations? In other words, is the dissociation point of the depth element of a visual impression within the conscious region, or is it infra-conscious? For me, if I can trust my powers of introspection, it is infra-conscious. Psychologically I cannot analyse the element of solidity in a given impression. I *know* perfectly well that it results from the synthetic combination of two retinal images. But it is for me a sensation which, though it results from a physiological synthesis, resists all my efforts after psychological dissociation. Others, however, affirm that it is possible by close analysis of visual impressions to distinguish between the two retinal images. For them the dissociation-point is within the conscious region. In any case the important point to notice is that the combination, however and wherever occurring (and it may physiologically occur in the lower brain-centres), does not merely produce a composite, and hence more or less blurred impression, but a clear-cut and definite impression, with a new quality, that of outness, distance, and solidity. If it be asked why the synthetic combination gives rise to this new quality in the psychical product, we must reply that we do not know. We know practically nothing concerning the ultimate "whys" of consciousness, though we know a little concerning the proximate "whys" and the "hows." We do not know why rays of a certain vibration frequency give the sensation red, while rays of another vibration frequency give the sensation green. We do not know why, when there are combined upon the retina the rays which give the sensation red with the rays which give the sensation green, we get a quite new sensation different from both, which we call yellow. So, for psychology, visual distance and solidity, the third dimensional space element, appears to be an ultimate element. We may explain, or attempt to explain, physiologically how it is generated, but why it takes this form we cannot say.

There are, however, other factors, which aid in giving to visual impressions their quality of distance, and to space, as presented to the eye, its element of depth. If we hold a pencil on our forefinger, about ten inches from our face, and focus our eyes alternately on this and on some distant object, we shall become aware, by the exercise of a little introspective attention, that there are present psychical elements other than those belonging to the visual sensations as such. These elements are sensations associated with certain movements of the eyes, or parts of the eyes, during the process of focussing. Physiology shows that these movements are of two kinds. First, there are movements of greater or less convergence of the two eyes, so as to make the image fall on the most sensitive area of the retina; and secondly, there are movements of accommodation within the eye by which, through a change in the curvature of the hinder surface of the lens, the image is rendered more clearly defined. It is probably impossible psychologically to distinguish between these two, though physiologically they are quite distinct. Moreover, these motor-sensations are seldom rendered focal except for purposes of psychological analysis. They are normally marginal, and are only dragged to the light of focal consciousness to satisfy the demands of scientific analysis. But I have written so far in vain if I have not made it clear that the influence of the subconscious margin, in determining our states of consciousness, is of extreme importance, and needs constant recognition.

When we see, then, a focal object set in a visual field, there are, apart from and in addition to visual sensations (that is to say, sense-data supplied by the retinas), motor-sensations (that is to say, sense-data due to movements of or in the eyes). And these latter, in correlation with the binocular factor, aid in giving to our impressions of sight their visual depth. They aid also in giving to objects their solidity. For when we take in the form or figure of any object, our eyes wander over its several parts with varying convergence and accommodation for distance. And it must be remembered that as, during our waking hours, we look hither and thither, and comprise now this, now that, and now the other object in our glance, we proceed by successive gradations, the motor impulses forming a continuous series associated with constantly changing adjustments. This continuity of the motor series is probably not a little helpful in giving rapidity and exactness to our visual localization.

It appears, therefore, that the depth element in visual impressions, as given through the combined impulses from two retinas, is correlated with motor-sensations of convergence and adjustment. We have now to see that the extension element, as given in the grouped impulses from each retina, is likewise correlated with other motor-sensations. To this end, let us begin by noting what takes place when we observe a moving object in the field of vision. We are watching, we will suppose, a game of billiards. Again and again our eyes follow the balls in their movements across the table. We are at most but dimly subconscious of our own motor-sensations as our eyes follow the movements of the billiard balls. It is the object itself as moving on which in each case attention is fixed. The ball is in the focus of vision, and as it moves the surrounding margin is gradually changing. Often two balls are moving at once. We fix our attention on one of them, and follow it with our eyes; but we are aware of the movements of the other as changes in the margin of vision; and we can leap with our eyes from one ball to the other with almost infallible accuracy. We can, if we like, while white and red are moving, fix our eyes on "spot," which is at rest. When the movements are entirely in the margin of the visual field, our eyes are still; there are no motor-sensations, but we have to exercise some control to prevent our eyes going off after one ball or the other, so strong is the tendency of our eyes to follow the moving object. So far as vision is concerned, therefore, movements are gradual changes in the relation of the focus to the margin, or relative changes within the margin, or both. And these changes are usually correlated with motor-sensations due to the movements of the eyes.

Now let us take a visual scene in which there is no movement. The game of billiards is over; the players have departed; only the psychologist remains. He fixes his eye on the red, and makes it focal in a marginal setting of the table with white and "spot" and a cue leaning at one corner. This catches his attention and away fly his eyes to render it focal. The whole scene shifts, and assumes new relations to a new focus. There has been no movement in the external factors of the scene, but there has been a complete rearrangement of the scene for consciousness; there is a new focus to a different margin. In that margin the large picture on the further wall catches his attention. It is a fair work of art with its focal point or centre of grouping sufficiently well defined. Our psychologist cannot, however, grasp the full work of the picture at a single glance. His eye wanders over it to embrace in succession the various details. He successively focusses all the essential features, and each successive focussing gives a new psychological impression, a new focus with a new margin. Then he returns to the centre of grouping, and begins to take in the full effect of the work of art as a whole. He grasps the general effect as a complex impression. For in any visual field there is a focal impression to which the rest of the field is marginal; and within the complex impression there is a centre of grouping to which the rest of the impression is relatively marginal. If, as we look at a picture, there is any subsidiary detail which catches the attention as not yet mastered, away go the eyes to this point to make it for the moment focal, and only when we feel that we have got its value for the total effect do we return again to the centre of grouping. In looking, therefore, at an unchanging scene there is abundant employment of eye movements, and we seem amply justified in saying that the relation of the focus to the margin (that is, the extension element) in any given visual scene is closely correlated with the eye movements necessary to bring any desired part of the margin into focus. The visual field is given in direct visual experience as extended and, when sensations of accommodation are added to the other motor-sensations, as having depth. There is not only a focus but a margin. And the relation of the one to the other is correlated with and suggestive of motor-sensations due to movements of and in the eyes.

Both the visual impressions and the correlated motor-sensations are presentative, they take their origin in impulses due to excitations of the afferent nerves. The visual field is, however, partly thrown into three-dimensional form through suggestion and association. The relative intensities, shading, shadows, the appearance due to haze, the relative blurring of objects in the margin, all these are suggestive of the distance element. As visual impulses they are, of course, presentative; but it is their suggestive or representative force that gives them value in this respect. The painter has to rely on such suggestions for giving the distance element or depth to his picture. But when we look at the picture, we feel that the depth, admirably as it may be represented, is suggested and not actually presented.

In the case of touch, as with vision, the extension which is given in tactile experience is closely correlated with motor-sensations. Suppose that we are feeling for an object in the dark, and that our wrist is brought into contact with it or with another object. There is not only a sensation of touch, but the impression is localized as originating at the wrist. Under such circumstances we at once move the hand and arm in such a way as to bring the fingers to bear upon the object; and the localization of the touch is correlated with the motor-sensations thus arising. Or suppose there is some source of irritation, say, at the back of the neck. At once the hand is carried to the spot to ascertain the nature of the irritant and to remove it. And here again the localization of the spot where the stimulus is applied is correlated with motor-sensations due to the movement of the hand to the spot. The correlation is by no means perfect in the absence of sight. If we get some one to press gently with the point of a pencil on our thigh or calf or upper arm, and then, with shut eyes, try and place our finger on the exact spot we shall probably miss it by an inch or so. Practice, however, rapidly and to a considerable degree improves the accuracy of the correlation and associated localization.

And not only is there a correlation between touch-sensations and appropriate motor-sensations, there is a close correlation between tactual and visual impressions. A friend has left on my table a fossil for identification. No sooner do my eyes fall upon it than my hand is reached forth to pick it up. Such association of visual and tactual experience is a very frequent and constant one from our earliest days. The child is repeatedly stretching out its hands, and we may observe how often the eyes follow the hands in their movements. In the first months of life there is ample opportunity for the establishment of close relationship between movements of the hands and fingers and movements of the eyes, and of both with changes in the visual field. When the child begins to crawl and move about, there is further opportunity for the establishment of correlations between these more extended movements and visual distance. Distance deepens to tactual experience, and new and most valuable correlations are formed between these new acquisitions and previous visual experience. And as the months go by there is a constant and progressive correlation of visual and tactual experience.

The exact mode of the establishment of this correlation is unfortunately left to conjecture. We have no recollection, or scarcely any recollection, of all that occurs in the building up of our psychical experience during the first two years of life. The probabilities are, however, that the psychical correlations are founded upon physiological co-ordinations which are inherited. This is a somewhat important point which needs a little explanation. By saying that the psychical correlations are founded upon physiological co-ordinations which are inherited, I mean that in the individual development the latter are prior to the former, the physiological co-ordinations to the psychological correlations. Consciousness, if I may so say, does not come into possession of a dead organism and then begin to pull the strings and make it work and live. Nor does consciousness bring with it any inherited knowledge of the world and what it is, of the body and how it works, or of the relation of one to the other. It is very doubtful whether conscious experience is or can be inherited. For some time consciousness is but a spectator, and, could he but remember, probably a dazed and confused spectator, of the physiological and organic reflexes of the body with a part of which it is closely if not inseparably associated. It is only gradually that consciousness begins to have some power of guidance and control over the organic mechanism. At first the organism is merely a marvellous physiological automaton, the nature and mode of working of which the nascent consciousness has to learn by experience.

A strong stimulus in a marginal part of the retina probably causes in the young child an organic reflex, by which those muscles are called into play which are fitted to turn the eyes, so as to make this stimulus focal. The co-ordinated movements of the eyes, their convergence and accommodation are almost certainly physiological, and due perhaps to infra-conscious impulses carried inwards by afferent nerves to the lower brain centres. But this organic reflex, though not in the first instance the result of conscious experience, is nevertheless accompanied by consciousness, and thus contributes to the upbuilding of experience. There is no conscious guidance in the initial stages, but conscious data are being provided for such guidance in the future. The stimulus of a bright object may also initiate, probably as a further organic reflex, the stretching forth of the hand towards it, while the contact of the object with the palm certainly gives rise to a grasping reflex. These reflexes, too, though not due to conscious guidance, are nevertheless accompanied by consciousness; and associations are thus established between the results of visual experience, and the results of motor and tactual experience.

How the physical data afforded to the infant by the early use of its sense-organs, and by the motor-sensations which accompany inherited reflexes, are organised into a consciousness which embraces them all in a complex unity we can but conjecture. It is not difficult, indeed, to build up a conjectural psychical baby, but it is equally easy to show how conjectural that psychical baby is. Of this however we may be tolerably sure, that it is through the grouping of the diverse data * and the correlation of the testimony

* We have some evidence concerning the correlation of newly acquired data in the case of those who have received sight as the result of a surgical operation. A world hitherto interpreted mainly through touch has to be interpreted anew for vision. It is said that to such individuals objects seem to be in contact with, or in close proximity to, the eyes; and that solid bodies, such as the cube or sphere, are taken to be flat surfaces. Even the human face seems to be a flat plane. One must be careful not to build too much on abnormal cases.

of the senses that the process is somehow effected. Consciousness is essentially a synthetic unity, and perhaps in this synthesis we may see a subjective aspect of that universal synthetic tendency which we discern in diverse forms throughout the objective world of nature,—a synthetic tendency which is seen alike in the genesis of a raindrop, of a crystal, and of the solar system; in the exquisite structure of the frustule of a diatom, in the form and brilliancy of a humming-bird, and in the silken gold of a maiden's hair. Many thinkers are unable to conceive how the synthesis of consciousness can be effected in the absence of a supernatural ego which effects the synthesis. For them the ego or individual personality is not the product of, but the supernatural producer of, the synthesis. But we are not wont to regard the genesis of a complex crystal as due to any individual act of supernatural synthesis, but rather as the outcome of a tendency which permeates inorganic nature. And in like manner we may regard the selective synthesis which gives rise to consciousness as due to a tendency which permeates the whole realm of psychical existence.

We are here however approaching a philosophical question which we cannot now discuss. It is sufficient in this place to note that the consciousness of the infant involves the extensive correlation of psychical data. What we call the extension and externality or outness of objects involves such correlation. The mere reception of stimuli on an extended surface does not suffice to give rise to the perception of extension. The difference between high and

---

The apparatus for convergence and accommodation, so long unused, could scarcely be expected to supply normal data to consciousness. The requisite stimuli for organic responses having been so long withheld, the apparatus for physiological response might well for some time be out of gear, and would thus supply data different from those which the healthy infant receives.

low musical tones is probably due to the stimulation of different parts of an extended surface, but they carry with them no extension element in consciousness. There is here no correlation, such as there is between visual and tactual extension. But though such organized correlation may be regarded as essential to the selective synthesis of consciousness, yet, I must repeat, the exact mode of its development and the stages of correlation are, and are likely to remain, conjectural.

Let us now summarize the correlations which we have briefly considered as contributing to the synthesis of our visual experience. There is, first of all, a correlation of the testimony of the two retinas, and here the correlation is probably infra-conscious; secondly, and concurrently, there is the correlation of visual sensations proper with the motor-sensations of accommodation; thirdly, there is a correlation of the changes which occur in the retinal field with the motor-sensations which arise from the movements of the eyes in their sockets. Co-ordinate and concurrent with this organization of a visual field through the correlation of retinal sensations with motor-sensations, is the organization of a tactual field through the correlation of sensations of touch with those motor-sensations which are subservient to tactual experience. And throughout all the stages of the organization of these two fields there are cross correlations, by which the experience gained in the visual field is brought into relation with experience gained in the tactual field. So that for normally constituted individuals experience is never gained separately in the visual field, and in the tactual field; but through constant correlation, the one with the other, the two fields coalesce in a common field of practical experience. Even so we have not nearly exhausted the many modes of correlation which enter into the fabric of our common work-a-day consciousness. There is an auditory field, an olfactory field, and a field of temperature. There is too a motor field, in which the movements of the limbs, head, and body as a whole are correlated. All of these, though of distinctly subordinate importance, are more or less definitely correlated with the fields of vision and of touch, and contribute to the orderly complexity of practical experience; and all are knit into an harmonious whole by that synthetic activity, in the absence of which there would be no states of consciousness at all.

And this complex process of correlation, the result of a natural and inherent synthetic activity, involves both presentative and representative elements. In the visual impression itself there is a synthesis of presentative impulses, and on this there rapidly follows a further synthesis, involving suggested representative elements. For the original presentative impulses, and the sensations to which they give rise, form nuclei, around which representative elements rapidly cluster and combine to form synthetic products of yet greater complexity. Let us steadily bear in mind that states of consciousness as we know them are wonderfully complex, with focal and marginal constituents, and with presentative and representative elements. We may in our analysis disentangle some of the strands of the many-hued, yet harmonious, tapestry of consciousness; but let us not forget that, as we know it in practical experience, the tapestry itself is a synthetic unity of marvellous and exquisitely intricate workmanship.

CHAPTER X.

THE SENSE-EXPERIENCE OF ANIMALS.

THE sense-experience, or consentience, as it has been termed, which results from the correlation of sense-data, forms the foundation of our psychical life; and it can hardly be questioned that it forms the foundation of the psychical life of animals. There is scarcely anything that has been said in the last two chapters which we may not fairly infer to apply in principle to those vertebrate animals, and especially the higher vertebrates, to which I have thought it best to confine our attention in this work. But when we go beyond these limits,—when, for example, we attempt to interpret the psychology of insects,—we find ourselves in face of serious difficulties, the nature of which may be indicated in passing, that my reason for restricting our zoological field may be made clearer.

Throughout our considerations the leading position taken by visual impressions has been obvious. Our world of practical experience is largely interpreted in visual terms. The retina of the vertebrate eye is so constructed that there is a small central area of maximum delicacy, the image or portion of an image which falls within this being for us, and we may infer for other vertebrates, focal. In other vertebrates as in us there is a mechanism for accommodation; while some other vertebrates, though only some, like ourselves use the two eyes in binocular vision. In all the vertebrates with which we are dealing, there are muscles which allow of delicately-graded movements of the eyes in their sockets. Hence we may infer that what we learnt concerning the correlations in vision, and the correlations of movements in the visual field with the motor-sensations which accompany the movements of the eyes in their sockets, holds good for vertebrates as for men. But when we pass to the insect,—the bee, for example,—there are two kinds of eyes, the small simple ocelli, and the pair of large many-facetted eyes. There is no mechanism for accommodation in either; no movements of the eyes in their sockets in either. It is not improbable that the ocelli serve mainly the purpose of directing the insect to a glimmer of light, the opening of the nest, for example; while the method of vision in the many-facetted eyes, the so-called mosaic vision, is quite different from anything of which we have or can have experience. In the visual experience of the bee, motor-sensations, special to the act of vision, can take no part. But we have seen that these motor-sensations in association and correlation with retinal sensations make *our* visual experience what it is. Does it not, therefore, logically follow that the visual experience of the bee must differ so widely from our own, that our wisest course is an honest confession of ignorance? Is not the bee in this respect like the kitchen clock; and are not we chronometers which only know the nature of their own insides? How little we know about the auditory and olfactory organs of insects; and what can we know at all about their auditory and olfactory experience? What know we concerning the tactile and other antennary sensations of the bee? What concerning the motor-sensations in that delicate organ? Still more, what concerning the synthetic correlation of the data of antennary senses and of antennary motor-sensations? In us the motor-sensations in the limbs are largely due to the movements of joints in their sockets; to some extent due to tensions in a soft and delicate skin. But in the bee the skin forms a firm encrusting armour, except at the joints; the muscles of the limbs lie within the tubular armour, and the joints are very different from ours. In us, the membranous labyrinth and semicircular canals give sensations accompanying changes in the direction or amount of movement in the head and body generally. We do not know of anything of the sort in the bees. If then the tactual field must be somewhat different from ours, the visual field widely different, and the antennary field — not improbably the dominant factor in bee-experience—quite unknown, must not one infer that the nature of the sense-experience of this insect is a secret she keeps to herself, even if she be philosopher enough to fancy she has guessed it? And if sense-experience be the foundation of the psychical life, that upon which the superstructure of higher psychical faculty is based, how far are our conjectures concerning the bee-superstructure likely to be correct, if we are almost wholly ignorant of the nature of the bee-foundation? Therefore, I say, let us not abate one jot of our industry in the objective study of the habits and activities of the invertebrate animals; but let us not pretend to know anything but the vaguest generalities concerning their psychology.

To return to the vertebrates; it appears to me that with regard to them, or in any case the more highly organised among their number, we are justified in inferring that the foundation in sense-experience is of like nature to that in man. No doubt there are differences, and not unimportant differences, in detail. There can be no question, for example, that in the dog and the deer the olfactory field is of far greater practical importance than it is with us. I have lately made experiments with a fox-terrier to illustrate the way in which he practically utilizes the sense of scent. On the steep slope of a hill where limestone fragments from the rocks above have collected in considerable quantities among may-bushes, nut-trees, and other vegetation, I have watched him search for the stones I have pitched down for him to fetch. Selecting a spot where a little rising ground effectually prevented him from seeing where the stone fell, I have sent him after marked limestone fragments. He watches the initial direction in which the stone is flying, and then races off. After that he trusts almost entirely to scent, though I am inclined to think that the sound of the fall of the stone gives him some guidance as to distance. He then works the ground to and fro and backwards and forwards by scent; not by any means very systematically, for he often goes over the same ground again and again. When the marked stone has merely been handled by me, he apparently scents it as soon as he comes within about a foot of it. But when he has brought it up in his mouth, and I throw it down again, he finds it much sooner, and scents it at a distance of about four feet. At this distance also he appears to sense the direction of the stone, for he turns rapidly and pounces upon it. In one set of experiments I sent him down again and again after a marked stone, pitching it down a hundred feet or so among the bushes directly he returned with it. He thus brought it back thirteen times in twenty minutes in broad daylight, and eleven times in the same period on a moonlit night, the side of the hill being in shadow. So completely does he rely upon scent, that I cannot readily get him to fetch a stone by sight. If I flick a stone down with my stick on to a heap of talus well within his sight, so that I can, and he could, see where it fell, he goes smelling over the heap, but fails to bring the stone. But if I set a stone rolling down with my stick, so that he follows it as it rolls, then he follows it by sight and brings it to me. There can be little doubt that sense-data of olfactory origin, in many animals, take a more prominent position in the organization of sense-experience than they do in man.

In the case of those animals—like the rabbit, for example—where the eyes are so situated that they cannot combine in binocular vision, one must suppose that the image which falls upon the most sensitive area, or yellow spot, of *one* eye, suggests the focal impression; while that which falls on the similar spot in the other eye is marginal to its conscious consentience. In them the binocular factor in sensing distance must be absent, and distance must be sensed through the synthetic correlation of retinal impulses and motor impulses of monocular accommodation. This is how the man who has lost one eye senses visual distance, aided of course by suggestions of relative size, the effects of haze, and other such representative factors. It is certain that the binocular factor is not essential for the sensing of distance, for the chick, in which this factor is absent, soon strikes with great accuracy at a minute object. Still it is probably a great advantage. If you hold a chameleon on your finger near a fly, within striking distance with his long and flexible tongue, he never strikes till both eyes converge upon the insect. The tip of the tongue is practically shot out like a projectile, and the accuracy of correlation of sense-data, and co-ordination of motor impulses for the innervation of the muscles, are alike remarkable.

Nothing is more admirable than the skill of animals. One may watch by the hour with ever renewed delight the marvellously delicate adjustments involved in the sailing flight of sea-gulls. I once saw at the Cape a seabird chasing another that had secured a fish. Hard-pressed the smaller bird dropped its prey, and with a swoop the other caught it in the air and sailed off with its stolen meal. An American observer, Mr J. Lancaster, describes in the *American Naturalist** the following scene. "A fine specimen of the fish-hawk swooped on a fish, which soon left its native element and swung aloft in the bird's talons. The hawk began its homeward journey. But now a new comer appeared upon the scene. A black creature, which seemed all wings, dropped from above and confronted the hawk, which at once let go its prey and uttered a scream so

* For March 1886, quoted in *Nature*, vol. xxxiii. p. 520.

brimful of mortal terror as to excite one's pity. The hawk was not struck, and it made off in wild haste for shore. The intruder was a frigate-bird, which seized the dropped fish in its beak long ere the prey reached the water, and then, with a sweep of exquisite grace, on tense wings, fronting a mild breeze, the corsair was lifted half a mile into the air. A bite was taken from the fish by a wringing motion of the bird's head, which sent the carcass whirling. The morsel being swallowed, the bird, folding its wings tightly to its body, dropped swiftly after the fish, seized it, again swept upwards, and then the performance was repeated till the meal was over." Every motion of birds in their flight is full of exquisite skill. Watch the swirlings of the swifts in the summer air; hundreds of insects, invisible to you, are being seized in a flight that is estimated at not less than a hundred miles an hour. Many of my readers must have watched at the zoological gardens, the skill with which the darter or snake-bird (*Plotus*) pursues fish, and captures them beneath the water, in the large tank at the end of the fish-house, or the skill with which the sea-lion catches the morsels which the keeper throws him. I watched the other day the unfailing accuracy with which an Irish terrier followed, and invariably caught at full-speed, a ball which his master flung for him across the open down; and I admired it all the more because I had been teaching a fox-terrier to do this, and knew that such skill is not to be acquired in a day or a week. One of the most striking examples of combined intelligence and skill, and one which is rendered more striking by contrast, is the way in which a sheep-dog works the relatively unintelligent sheep on a steep and broken mountain-slope. Whether we regard the natural skill of wild animals, or the skill domesticated animals display under human guidance and for the service of man, we cannot fail to admire, if not to wonder. And since this skill is unquestionably a matter of sense-experience, it will be well to consider carefully its psychological aspects.

In the first place it should be noticed that although, as we shall see more clearly in the next two chapters, skill is displayed in virtue of the possession of an organic mechanism which is inherited, and although the performance of all the essential activities is founded on a basis of an innate automatism, yet all the accurate finish and final touches of perfection are the result of individual acquisition. And it should be further noticed that this individual acquisition is something more than and something other than the increased smoothness of working of a piece of automatic mechanism. A new machine works better after it has been running for a while; but the perfecting of animal skill is something more than this. I would ask anyone who is inclined to doubt this, and to regard animal activity as perfected automatism, to teach a dog to race after and catch a ball as it flies and bounds over a lawn. He will, if I mistake not, be forced to conclude that the individual acquisition of skill through practice is not merely an increase in the smoothness of working of the automatic mechanism, but involves a delicate power of control over the motor apparatus. If we call this perfected automatism, we should remember that the automaton is one which profits by experience. And it is well to bear in mind that the words "automaton" and "automatic" are used in two somewhat different senses. An automaton is defined as "a self-moving machine," and automatic as "having the power of moving itself." If now we lay stress on the *self-moving*, we have one sense in which the word is used. In this sense I am quite prepared to regard myself and animals as conscious automata. But if stress be laid on the *machine*, then the word automatic acquires the connotation of mechanical uniformity of action. In this sense I do not regard myself and animals as automata. It is not for me to decide between these two uses. I shall, however, myself in this book use the words "automaton" and "automatic" in the sense in which they imply mechanical uniformity of response. I shall speak in the next chapter of the just-hatched chick as a little automaton, meaning thereby that in virtue of its organic mechanism it responds like an accurately made and nicely contrived machine to certain stimuli. And I shall use the word "automatism" in antithesis to control. The automatic act may be *accompanied by* consciousness, but the controlled act is *guided by* consciousness. And I think there can be no question that the perfecting of an act of skill is under the guidance and control of consciousness.

There are two factors in the perfecting of skill,—the one is the nice correlation of sensory data; the other is the co-ordination of motor impulses or of muscular adjustments. The delicacy of this co-ordination, itself dependent on a like delicacy of correlation of sensory data, is seen in the products of human art, in etching, engraving, wood-carving, or violin playing. Nor is it only seen in the play of the more refined finger-muscles; it is seen also in what we are apt to regard as the coarser and grosser muscles of the trunk. The cricketer, the bicyclist, the gymnast; the angler, the archer, and the marksman at the rifle-butts; all these illustrate the pitch of accuracy which the co-ordination of many and differently situated muscles may reach. Let us take the rifleman for further illustration. At a range of 1000 yards, a deviation of the muzzle of the rifle by less than $\frac{1}{100}$ of an inch involves a deflection of the bullet when it hits the target of six inches from the centre of the bull. The rifleman, we will suppose, is lying down prone, supporting his elbow on the ground, and grasping the rifle, say half-way between the fulcrum, where the stock is held to the shoulder, and the free muzzle of the piece. A deviation of the muzzle by $\frac{1}{100}$ of an inch to left or right involves, therefore, a deviation of the hand by half this amount, or $\frac{1}{200}$ of an inch. But the hand moves around the axis of the arm, and is carried at the end of a lever consisting of the fore-arm from elbow to wrist, and this movement is effected, in the main, by muscles around the shoulder. A relatively small movement of the point of insertion of one of these muscles produces a relatively large movement of the hand and wrist. Hence there is certainly no exaggeration—nay rather, we are probably far within the mark—in saying that the skilful marksman has his motor co-ordination under control to within the 500th if not the 1000th part of an inch; and this in the play of muscles which would not generally be regarded as susceptible of the highest delicacy of such co-ordination. Nor is the accuracy of co-ordination more remarkable than the rapidity with which in some cases it is effected. Instance the violinist playing a rapid passage at sight, the swordsman whose life depends on accurate and rapid parry, or the cricketer at the wicket who has made all the co-ordinations necessary for cutting the ball in the second or so that elapses between its delivery and its reaching the crease. Or instance, in the animal world, the dog catching the flying ball, or the swift sweeping down on minute insects.

Such cases as those of the swordsman and the cricketer serve to bring out another feature in perfected skill, namely, the suggestive value, through the organization of experience, of the initial stages of the action, the effects of which are to be met by the skill, and the part performance of the answering action in advance. Shortly after the ball has left the hand of the bowler, the batsman is prepared to strike in a certain way. Of course, he is still on the alert; for the ball may shoot, and this will necessitate a rapid and special co-ordination. So, too, the swordsman knows from the first movement of his adversary's point what is likely to follow, and the antagonist he fears most is the master of newly devised methods. And so, too, when we watch two dogs fighting, or two well-matched cocks in a farm-yard, we see how every movement and every initial stage of attack are closely watched and call forth the answering movements, which are in turn not less closely watched by the antagonist.

Now I suppose we may take it for granted that the perfecting of skill under conscious guidance cannot be effected unless the net results of motor co-ordination come within the ken of consciousness. But, seeing how delicate is the co-ordination of motor impulses, which must often be very numerous, it is at first sight somewhat remarkable that it is seemingly only of the net results, scarcely at all, if at all, of the details, that consciousness is aware. Take, for example, the case of the rifleman, which we briefly considered. I doubt whether many marksmen could tell us, in the absence of experiment and observation specially directed to this end, whether the muscular adjustments, the net results of which he is consciously or subconsciously utilizing, occur mainly in the wrist, in the forearm, in the upper arm, or in the shoulder. Perhaps the last muscles he would think of particularizing are those around the shoulder-blade. But if he were wise enough to state only what he actually experiences, he would probably say, "I am only aware generally of certain effects upon the direction of the rifle; I am unaware of any of the details of the process." But though in practical experience we are unaware of these details, it is just these details which give accuracy to the general adjustment of the net results as a whole. And psychology must attempt to explain this seeming paradox.

Let us first notice what a boon it is to consciousness that it has not to deal with all the details. How many muscles are concerned in maintaining a man in the erect position when he is standing at ease I do not profess to know. But I question whether there is a physiologist living who could give an accurate diagrammatic representation of the relative intensity of the innervation of, let us say, a hundred of the most important of these muscles. And when a man, instead of quietly standing at ease, walks, or runs, or cuts a ball at cricket, the duly graduated intensities of the series of innervations involved are quite beyond our powers of descriptive analysis. And yet, what more easy than to walk, or run, or even make a decent cut at cricket? It is a great boon that all the subtle details are left to physiological co-ordination, while consciousness is free to deal only with the net results, or groups of co-ordinations as wholes; that a number of physiological impulses conspire to form a single motor-sensation.

And the method employed by consciousness is the method of trial and error. Take the case of the marksman. His object is to find that motor adjustment which shall produce certain effects in his visual field, so that, on a still day, the sights and the centre of the bull are in a right line. And when, by trial and the selection of the successful adjustment, he reaches the desired result, he has to exercise control over the motor adjustment so as to prevent its falling away from accuracy. That the net results of the motor adjustments are really felt, and accurately felt, seems to be conclusively shown by such an experiment as the following. Stand before a target with a pistol. Having taken a view of the target, shut the eyes, raise the pistol, and fire. You will perhaps be surprised at the comparative accuracy of your aim. You have, however, fired too high, and to the left. Fire again under similar conditions. You will find that you have corrected, probably over-corrected, your previous error. And this you could not possibly have done unless the net results of the motor adjustments had been represented in consciousness with very considerable accuracy.

We are perhaps apt to be misled by the fact that motor adjustments, except for physiological and psychological purposes, are seldom matters which we attempt to describe or explain to our neighbours. Through language we come to live in such a world of description and explanation, that we are in danger of fancying that what we do not deal with in these ways does not come into consciousness at all. Now it is one of the characteristic peculiarities of skill, that you cannot effectually describe or explain how the motor adjustments are to be made. Hence it comes that five minutes' demonstration is worth more than five hours' talking where the object is to impart skill. It is of comparatively little use to describe or explain how a skilled feat is to be accomplished; it is far more helpful to show how it is done. And when we do describe or explain, we do so, not in terms of the motor adjustments themselves, but of their visible effects. In teaching a boy to play billiards, we mingle some description with our demonstration; but what we describe is not the motor adjustments, but their effects. We say, "Strike your own ball here, moderately hard, and aim to hit the red on this chalk mark I made on it." He plays, and fails. We explain that the failure was due to his striking his own ball too hard, and the red too centrally and fully, and bid him try again. Gradually, by trial and error, he acquires the skill we would impart. But his attention has been fixed throughout on the effects of the motor adjustments, not on the motor adjustments themselves. He would find it exceedingly difficult to describe these motor adjustments as such. And he might even fancy, like some worthy people whom I have met, that they do not enter the conscious field at all. Experiment and observation directed to that end would soon, however, convince him, if he cared to look into the matter, that they are not only not unfelt, but that they are felt with an accuracy that is somewhat surprising. They are, however, habitually marginal in consciousness, and only for purposes of psychological analysis are dragged into the focus.

A good example of motor adjustments, which though marginal are certainly not unfelt in their net results, is afforded by the production of the voice in singing. And here no description nor explanation is of much service. We sound or sing a note to a child. At first the attempts to reproduce the same note are perhaps somewhat wild. The voice pitches now here now there, now too high and now too low. But after some attempts, and often some sliding, the note is caught, and is then held to with some steadiness, if the child have a musical ear. Gradually control is gained over the laryngeal apparatus, until the visual stimuli of a vocal score suggest with accuracy and rapidity a sequence of intervals, each of which involves a special motor adjustment. And these motor adjustments are certainly not unfelt. I do not mean merely that the whole process of learning to sing at sight would be inexplicable if they were unfelt; I mean that I personally feel them with great distinctness. If I am looking over some new songs in a music-shop, I quite clearly feel the motor adjustments necessary for singing the notes set down, though I do not murmur a sound. And if the song runs up beyond my vocal range, I have an uncomfortable strained sensation in the throat. For every note I can sing I have a separate motor-sensation; and this sensation, though for psychology an undecomposable element in consciousness, is physiologically the net result of a great number of impulses from the several parts of the larynx. And the acquisition of skill in singing at sight involves the nice correlation of these motor-sensations, with certain visual sensations due to stimuli from the vocal score, and certain auditory sensations which accompany the production of the voice. Of course this analysis is incomplete. Motor-sensations accompanying the breathing, and the position of the tongue, lips, and mouth organs, require delicate correlation with the laryngeal and auditory sensations; and no doubt there are further subsidiary correlations.

I have said so much concerning skill, because, in the correlations it involves, we have some of the most delicate and finished results of the selective synthesis of sense-experience or consentience. And I have said so much concerning the human aspects of skill because it is necessary, in accordance with the method of interpretation adopted in this work, to study the psychology of skill at first hand in order that we may infer what takes place in the minds of animals. I think there can be no question that the acquisition of perfected skill by the dog or the elephant, by the beaver, the swift, or the snake-bird, is effected in analogous ways to those by which it is effected in man, excepting in so far as in man there is a certain amount of guidance by description and explanation. This factor is certainly absent in animals. No one ever taught a dog to perform the simplest trick by describing or explaining how it is done. Animals have to rely first on the internal promptings to satisfy hunger or other emotional affections; secondly, on parental guidance; and thirdly, on the influence of what has been termed tradition—that is, the continuity of habit in a community. Animals which live in herds, packs, or flocks, are born into a community which perform certain actions in certain ways, and through imitation of these traditional proceedings they have a strong tendency to act in like manner. But the acquisition of skill, however prompted, is essentially the individual perfecting under conscious control, of activities, the basis of which is an inherited aptitude. This perfecting involves the correlation of sense-data, visual, auditory, olfactory, and so forth, on the one hand, and motor on the other hand; and it is furthermore inexplicable in the absence of motor-sensations which enter, if only marginally, into the field of conscious experience.

It is through skill, and the application of skill, that animals deal so successfully with their environment. For when skill has been acquired its application is based on association under the varied and varying conditions of sense-experience. The application is essentially a practical matter in accordance with the needs and requirements of a life that is full of vicissitudes. By the more intelligent animals every favourable association is utilized and becomes a factor in the further development of experience. In future chapters the nature and range of intelligent adaptation to circumstances will come under our notice and will receive fuller illustration. All such intelligent adaptation falls within the range of sense-experience, which deals with sense-data of all kinds, and correlates them for the purposes of practical guidance, thus enabling the animal to carry out its life-activities and to meet the varying exigencies of a complex environment.

In conclusion, it should be remembered that both in the delicacy of their sensory endowment and in the ability to deal with that environment by sense-experience, animals are probably in some respects distinctly in advance of man. Witness the delicacy of the sense of smell in some animals enabling them to do that which no man could do so well. I think it not at all improbable that their powers of rapid flight in the free medium of the air have induced in birds a delicacy and high specialization of the sense of direction of the movements of the body as a whole, of which we slow treaders of the ground can scarcely form any conception. And this perhaps is a factor in that most difficult and complex question,—one that involves a good deal more than instinct only,—I mean the migration of birds. Very possibly our own endowment, in this sense of direction of movement, is very degenerate; still more probably we civilized folk do not make much use of the sensory endowment that we have, relying rather on our *knowledge* through spatial perceptions and conceptions than on our native powers of sense-experience. Savages who in their daily life make more use of these native powers, and explorers who are led by circumstances to cultivate these powers, have a sense of direction that the city clerk has allowed to lie dormant and unused. He does not cultivate the delicate use of his membranous labyrinth, with its semi-circular canals. Animals do cultivate the use of this sense-organ, and probably elements contributed thereby enter into the organised and correlated field of sense-experience in a way at which we can but dimly guess.

CHAPTER XI.

AUTOMATISM AND CONTROL.

WE have already learnt that from all parts of the surface of the body,—from eye, ear, nose, and palate, from the muscles, joints, and internal viscera,—there run nerves—ingoing or afferent nerves—which are in direct connection with the spinal cord and the brain. These nerves are the channels along which impulses may be transmitted to the nerve-centres. Within these nerve-centres molecular disturbances are thereby caused, as the result of which further impulses are transmitted down other nerves—outgoing or efferent nerves—to muscles, which are thus stimulated to contraction; or to glands, which are stimulated to secretion. It is generally believed, as the result of many observations and experiments, that consciousness is not associated with the transmission of impulses along the nerves; but that when the molecular thrill reaches the brain, or some part of it, *there*, under appropriate conditions, consciousness emerges.

The group of structures consisting of (1) afferent nerve, with its sensory termination, (2) nerve-centre, and (3) efferent nerve, with its motor or other termination, is spoken of as the *nervous arc.* It is called into activity by a *stimulus* applied to the sensory termination of the afferent nerve, and the motor or other result of its activity is spoken of as the *response.* The response is always more or less complex, involving the balanced and nicely regulated action of several, often many, muscles. And it is clear that the balanced action of several muscles must be due to the transmission to them of separate impulses duly graduated in intensity. The graduation of the outgoing impulses, as the result of the receipt of incoming impulses, is effected in and by the nerve-centre, and is spoken of as *co-ordination*; and this co-ordination may or may not be accompanied by conscious *control*. It is with the nature and the relations of co-ordination and control that we have to deal in this chapter. But we must deal with them in mere outline, and, so to speak, diagrammatically. The physiology of the question is difficult and full of technicalities; and to a large extent it is external to, though closely bordering on, psychology. Unfortunately, too, the exact relations of physiological processes to the states of consciousness which we experience are still obscure and in large degree conjectural. I can only give the view which appears to me the most probable and that which is in best accord with known facts, not a tithe of which can be here adduced.

There can be little or no question that what one may term the groundwork of co-ordination is a matter of inherited organic structure in the nerve-centres. Hence, not infrequently, co-ordinated responses to stimuli are tolerably perfect at or very soon after birth. In all the higher mammals the exceedingly complex series of motor operations concerned in sucking and swallowing is at birth performed with sufficient exactitude to ensure the continuance of life. The little pig, or the baby hippopotamus, can very soon run about with surprising assurance. The human infant, though it is so much more helpless, clings to your fingers with tenacious and well-regulated grip, and will hang suspended for a couple of minutes. Even the sneeze with which it greets the world, its cries, and the seemingly aimless movements of its limbs involve co-ordination more or less perfect.

It is, however, especially in those young birds which, like the chick, undergo prolonged development within the egg, and are hatched in a fairly advanced state of organic development, that inherited co-ordination of response is shown in its most marked form. If a chick, hatched out in an incubator, be kept secluded from the world for twenty-four hours or so and then be placed under observation, it will "cheep," walk, and peck with fair but not complete accuracy at crumbs or bits of white of egg. If a fly with clipped wings be placed before it the chick will follow and peck at it, catching it perhaps at the sixth or seventh peck and swallowing it. If water be placed before the chick in a shallow tin it will perhaps take no notice; but if it be induced to peck at something in the water, it will lift its head in the peculiar way fowls have when they drink. The stimulus of water in the beak gives rise to a very special co-ordinated response involving a great number of muscles. Now in this and other such actions, *the first time each is performed*, the chick would seem to be a little automaton,—no doubt a conscious automaton in so far as the performance of the activity in question is accompanied by consciousness, but none the less an automaton in the sense that it performs the action merely in virtue of its living organic structure, because it is a piece of beautifully delicate, *going* mechanism. The action is automatic as opposed to controlled. For it does not seem possible for control to be exercised over an activity in the absence of any experience of the nature of that activity. It may indeed be suggested that ancestral experience of the performance of such activities may be inherited by the chick. But I hold that the hypothesis of inherited experience is unsupported by evidence and unnecessary, since the facts may be more simply explained on the view I am here advocating. What is inherited is organic structure fitting the organism for the automatic performance of complex activities, the co-ordination of which is moderately accurate from the first. The initial performance of the activity is, however, accompanied by consciousness;

and here we have the primary basis in experience for the subsequent control of the activity in question.

For no one can watch the development of chicks or other young animals during the early days of their life, without obtaining abundant evidence of control. Chicks, for example, at first peck indiscriminately at almost everything that catches their eye, notably at their own excrement. But this is bitter or unpleasant to them. And you may watch the stages of gradual avoidance of pecking at this material, ranging from mere hesitation to complete neglect. Ladybirds are at first picked up, and then dropped with some comical wiping of the bill on the ground; very soon they are left quite untouched, and apparently unnoticed. These, though simple cases, are examples of activities under control as the result of experience. And while nasty things are thus avoided, nice things are seized with increasing avidity and eagerness. Here, in fact, we have in simple expression that which is the essential feature of control, the inhibition of activities which are distasteful and associated with discomfort, and the conscious reinforcement of activities which are pleasurable and lead to satisfactory results.

The question now arises: Can we distinguish between nerve-centres whose function is the automatic co-ordination of activities, and other nerve-centres which are associated with the function of control? With a view to finding an answer to this question, let us run briefly over some of the chief features of the anatomical structure of the central nervous system. In the first place, there is the spinal cord, lying within and protected by the arches of the vertebral column. Communicating with it there are a number of spinal nerves, each of which just before joining the cord divides into two so-called roots, of which one, that nearer the mid-line of the back, contains the afferent fibres; while the other, or ventral root, is composed of efferent fibres. The cord itself consists of right and left halves, connected each with the other by narrow central bridges of nervous matter. The essential structures in each half are,—First, an internal nervous tangle of delicate fibrils, some of which are connected with conspicuous nerve-cells, the tangle and the cells constituting the grey matter. This is to be regarded as a series of connected co-ordinating centres, but whether the conspicuous cells are mainly instrumental in co-ordination, or whether their function is chiefly that of maintaining in due order the nutrition of the nerve-fibres and fibrils, is an unsettled point. Secondly, external to the grey matter are longitudinal tracts of fibres running up and down the cord, one of which lying on the dorsal side may, for reasons which will shortly appear, be particularized. It is termed the pyramidal tract. This and other tracts constitute the so-called white matter of the cord. Thirdly, there are the bridges of nervous matter connecting the two halves of the spinal cord.

Traced upwards, the cord expands at the base of the skull to form the lowest part of the brain, termed the bulb or medulla oblongata. Here the internal grey matter, of increased size, and, external to it, the white matter, are still recognizable; the pyramidal tract is conspicuous, and to a large extent crosses over, right to left and left to right. Above the bulb, lying between it and the large cerebral hemispheres, is a complex series of nerve-centres with tracts of fibrous white matter, which it must suffice for us to group together as the "lower brain centres." Here and in the bulb are the centres in association with the optic nerve and with the cranial nerves coming from tongue, palate, ear, and the skin of the face; and going to the muscles of eye, tongue, and jaws, and to the larynx, lungs, heart, and stomach. But these only form a small proportion of the grey matter of these centres. A far larger proportion is composed of a nervous tangle, concerned, as we shall presently see, with the co-ordination of activities, and receiving from all parts of the cord ascending or afferent impulses by a special tract, the so-called "median posterior tract." Right through these lower parts of the brain run strands of fibres continuous with those to which allusion has more than once been made as constituting the pyramidal tract.

The highest brain-centres are the cerebral hemispheres. These, as one of our greatest living physiologists has said, seem to stand apart from the rest of the brain. They are connected by tracts of fibres with the lower brain-centres, including the cerebellum; while the cortex, consisting of grey matter formed of a nervous tangle with nerve cells, is directly connected with the before-mentioned pyramidal tract. This tract consists entirely of outgoing or motor fibres, and by its means the cerebral cortex is in direct connection with the centres associated with those cranial nerves which are motor in function, and with the centres of the spinal nerves throughout the whole length of the cord. These facts, then, have especially to be noticed with regard to the cerebral hemispheres; first, that they stand to some extent apart from the rest of the central nervous system; and, secondly, that though there seems to be no direct connection between the sensory nerves and the cerebral cortex (such connection being indirect through the intervention of lower brain-centres), there is, by means of the pyramidal tract, a direct connection between the cerebral cortex and the motor-centres of both cranial and spinal nerves. In view of these facts, the following hypothesis suggests itself:—that the spinal cord and lower brain-centres afford the mechanism of co-ordination; that from this mechanism impulses pass upwards to the cerebral hemispheres which constitute the organic mechanism for control; and that this control is brought to bear directly on the motor centres by means of the pyramidal tract. The diagrammatic representation of the nervous system given in Fig. 11 is drawn in accordance with this view. The cerebral hemispheres at the top of the figure are separated off from the rest of the nervous system, and the pyramidal tract is separated from the rest of the spinal cord. It should be noted that this tract is relatively most marked in man; it is less well developed in the monkey, still less in the dog, and is scarcely if at all recognizable in the lower vertebrates. It is probable, however, that in all cases there is some direct motor communication between the cerebral hemispheres, or mechanism of control, and the several motor centres of the mechanism of co-ordination.

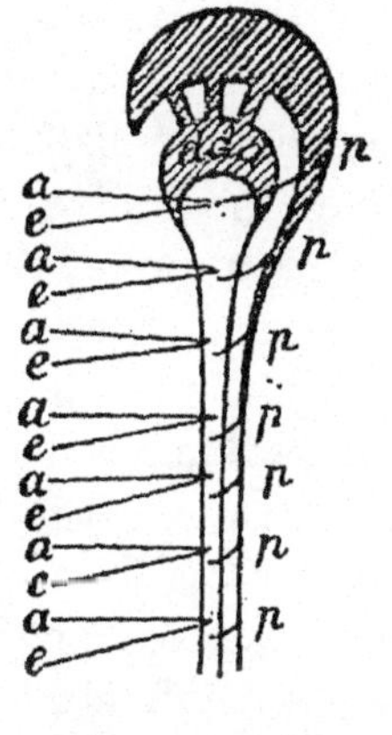

FIG. 11.

a *Afferent nerves.* e *Efferent nerves. Above* d d d *are the tracts of communication between the lower brain-centres below, and the cerebral hemispheres above. Opposite* p p *are the fibres by which impulses are carried from the cerebral hemispheres through the pyramidal tract to the motor centres in the base of the brain or the spinal cord.*

We must now briefly consider how far known facts of observation tend to bear out this hypothesis. In the frog, the cerebral hemispheres may be removed without killing the animal, which may be kept alive for a prolonged period after the operation. Such a frog may perform all the necessary and usual bodily movements almost or quite as well as an unmutilated animal. The removal of the cerebral hemispheres does not seem to have materially modified the mechanism of automatic co-ordination. The frog will eat when food is placed in its mouth, will even catch and swallow flies which come into its neighbourhood. If laid on its back it will, after a few struggles, regain its natural position. If stimulated, it will crawl, and leap, and avoid conspicuous obstacles. If thrown into water it will swim vigorously and for a long period, and will crawl out on a piece of wood or on the sides of the pool. Placed on a board, it will sit up in the usual attitude; but when the board is so tilted that the body is thrown out of equilibrium, the frog will crawl up so as to reach a secure position. By constantly moving the board, the animal may be kept on the move for an indefinite period. If its flanks be gently stroked the frog will croak, answering to this stimulus with mechanical regularity. In all respects the frog appears to be a responsive automaton, with definite and accurate powers of co-ordination. But all volitional control of action appears to have been removed. The animal seems quite incapable of profiting by experience. If left to itself, it will remain squatting on the same spot until it dies.

In warm-blooded animals the effects of the removal of the cerebral hemispheres are more serious, and it is less easy to keep them alive. In birds, however, when they survive the shock, a drowsy sleepy condition follows the operation. They assume a natural position, and only move therefrom on the application of a stimulus. This sleepiness may pass by, and the bird may walk about restlessly and aimlessly, just as the frog swims when it is placed in water. A pigeon, so deprived of its cerebral hemispheres, will turn out of her course to avoid an obstacle; but it makes no difference whether the obstacle is an inanimate body, a cat, a dog, a bird of prey, or another individual of the same species. She is sensitive to sounds as such, but it matters not whether the sound be the cooing of a mate, the rattling of peas, or the call-whistle which suggests feeding-time to its uninjured companions. The male pays no attention to the hen; the mother takes no notice of her young. On the whole, then, the observations which have been made seem to justify the view that by the removal of the cerebral hemispheres the bird has been deprived, not of its power of co-ordination, but of its power of voluntary control. And the observations that have been made on mammals, during the short time that they survive the operation, do not tend to negative but rather to support the view that the loss of the cerebral hemispheres involves loss of voluntary control but not loss of adapted co-ordination. Automatism remains, but of ability to profit by experience there is little or no evidence.

If we thus regard the cerebral hemispheres as constituting the organic mechanism of control, and the pyramidal tract as the channel by means of which control is brought to bear on lower centres of co-ordination, we must also unquestionably regard the apparatus as delicately differentiated, and the differentiations as transmitted by inheritance. Much admirable physiological research has been of late years directed to the mapping out of the motor control centres in the cerebral hemispheres. These are often spoken of as motor centres, but the use of the phrase control centres more accurately suggests their true function. It is noteworthy that the stimulation of the appropriate centres calls into play not merely muscular contractions but co-ordinated motor activities. And there can be little question that each centre, either through different nerve-fibres or more probably through the same fibres, has a two-fold function,—that of enforcement of the activity, and that of inhibition or checking of the activity. How this inhibition is effected is a difficult physiological problem, which does not concern us here. Nor do the details of the mapping of the centres in the cerebral cortex concern us. This is a matter of brain-physiology, not of psychology. There is, however, one physiological, or rather anatomical, observation which, if it be confirmed, is of great interest from the psychological point of view. It is stated that in the brain of the newly-born infant, while the fibres running up from the lower brain-centres to the cerebral hemispheres and those running from the cortex to the pyramidal tract are already developed, the "association fibres"—that is to say the fibres running in the cerebral hemispheres from centre to centre—are not developed but are established in the individual after the brain has entered upon its duties and its functional activity. This anatomical observation accords with the psychological view that the association of ideas is a matter of individual establishment, and is not, at any rate in its details, a matter of inheritance.

On the hypothesis which is here adopted, the animal which is deprived of its cerebral hemispheres is a mere automaton; and the question arises whether, under these abnormal conditions, it is a conscious automaton. The answer to this question is, and is likely to remain, wholly conjectural. I shall assume that it is an unconscious automaton, and adopt the hypothesis that consciousness is associated with the molecular disturbances which take place in the cerebral hemispheres. My reason for taking this view is that, as it seems to me, the primary aim, object, and purpose of consciousness is control. Consciousness in a mere automaton is a useless and unnecessary epi-phenomenon. As will be seen in a later chapter, however, I am very far from denying that some dim foreshadowing of that which we know as consciousness is an accompaniment of the functional activity of *all* nerve-centres. But I think that we may well regard this as lying in the infra-conscious region of our psychical curve, and therefore as extra-marginal to the field of our ordinary consciousness. I assume then that in us and in the higher animals (vertebrates) consciousness is associated with the functional activity of the cerebral hemispheres.

Let us now go back to our little chick that pecks more or less automatically at some white unpleasant substance, but, finding it distasteful, checks after a few trials the motor activities involved in pecking and leaves the substance alone. And let us endeavour to represent in a diagram the manner in which such control may be brought to bear. It must be remembered that this is a diagram and nothing more, and if at first sight it looks somewhat complex we must remember that the probabilities are that it errs, among other things, in being almost absurdly too simple. In the diagram the cerebral hemispheres are as before represented, for purposes of illustration, as separated from the lower brain-centres which lie beneath, and the pyramidal tract is also separated off from the rest of the nervous system. It is so drawn to emphasize the hypothesis that the mechanism of control is distinct from the mechanism of automatic co-ordination. The first stage of the process is the reception of a retinal stimulus. This gives rise to an afferent impulse by which the centres for the co-ordination of the activity of pecking are thrown into activity, as the result of which the efferent impulses for that action leave the cord at the lower part of the diagram. This is the primary result of the stimulus. But there is a secondary result. The impulse not only sets in motion the co-ordinating apparatus, but (probably through the fibres known as the "optic radiation") throws into activity the visual centres V

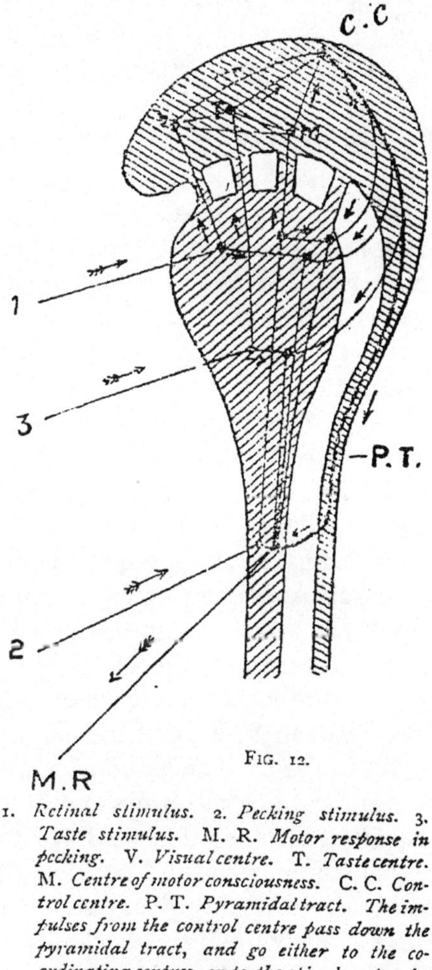

FIG. 12.

1. *Retinal stimulus.* 2. *Pecking stimulus.* 3. *Taste stimulus.* M. R. *Motor response in pecking.* V. *Visual centre.* T. *Taste centre.* M. *Centre of motor consciousness.* C. C. *Control centre.* P. T. *Pyramidal tract. The impulses from the control centre pass down the pyramidal tract, and go either to the co-ordinating centres, or to the spinal centre for motor response, or to both.*

in the cerebral hemispheres. So much for the results of the retinal stimulus. The action of pecking now takes place. But this activity stimulates a number of afferent fibres, giving rise to what is termed in the diagram a pecking stimulus, and the impulses thus started probably again have two results. In the first place, they may have a direct influence on the centres for automatic co-ordination; and secondly, they throw into activity the cerebral centres M, which are associated with motor consciousness, or the consciousness of the action of pecking which is taking place. Now there probably already exist in the hemispheres inherited and well-established connections between the centres of motor consciousness and the control centres of that particular activity. Hence the control centre is thrown into activity, and through the pyramidal tract either enforces or inhibits the continuance of the activity. We have now three centres in the cerebral hemispheres thrown into activity in rapid succession, the visual centres V, the centres of motor consciousness M, and the control centre. And here we must call to our aid the fact, which psychologically is unquestionable, of association. The states of consciousness concomitant with the activities of V, of M, and of the control centres, if such there be, thereafter tend to be suggested in the order of their primary occurrence. And this we must represent on our diagram by connecting lines between V, M, and the control centre.

But in the case under consideration, the action of pecking causes the chick to seize something that has an unpleasant and bitter taste, so that we have following rapidly upon the retinal stimulus and the pecking stimulus a taste stimulus. This again probably gives rise to impulses which first take effect directly upon the mechanism of automatic co-ordination, but which also throw into activity the sensory centres for taste T in the cerebral hemispheres. The sequence is so rapid that probably all four centres in the hemispheres, the visua centres, the taste centres, the centres of motor consciousness, and the control centre, are all in a state of molecular disturbance at the same time, and thus all become associated each with the other. But the consciousness which is concomitant with the activity of T is unpleasant and distasteful, and the effect of T upon the control centre will therefore be inhibitive. And even supposing that the effects of V and of M upon the control centre are slightly augmentative, and tend in some degree to enforce the activity, still the strongly inhibitive effect of T will prevail and the activity will be checked.

Such, in brief, is the way in which, in a very simple case, control *may perhaps be* brought to bear upon the mechanism for automatic co-ordination. All such schemes must, however, be regarded as the merest diagrams, and of little value save as showing the kind of way in which control is anatomically or physiologically possible. There must, of course, be nothing in them which contradicts known facts, anatomical, physiological, or psychological. And this, I venture to hope, is the case with the scheme here given.

It is well to restrict the term *reflex action* to that which is performed through the instrumentality of the mechanism for automatic co-ordination, and of this alone. Thus the first peck resulting from a retinal stimulus is a primary reflex action. I am aware that some eminent physiological psychologists are wont to urge that all cerebral activity "partakes of the nature of" or "may be reduced to the type of" reflex action. And if anything which involves a nervous circuit be termed reflex action, this is logical enough. But on the view I am here advocating it is well to restrict the term reflex action to that which is a matter of animal automatism, and thus to mark and emphasize the distinction between this and the higher process of control. It is probable, however, that control serves to modify and impress new characters on the automatic responses of the co-ordinating mechanism. Control involves, and probably always involves, a double circuit,—the lower reflex circuit and the higher control circuit. When during the life, and especially the early life, of the individual, the lower reflex circuit is constantly aided and enforced, or at any rate uniformly permitted, by the control circuit, it will become thereby the more firmly established; where it is inhibited, it will tend by disuse to disappear as a reflex; and where it is modified, certain elements of the activity being reinforced and others checked, it will tend to persist in this modified form. Control may then be withdrawn, and the co-ordinating apparatus left to carry on the activity in its modified form. Thus are habits formed, and a whole class of activities known as secondary reflex actions are established. Such habits will carry with them the purposive nature impressed upon them by the control under which they have taken form. Suppose, then, a pigeon, in which such habits have been established, be the subject of the operation of the removal of the cerebral hemispheres, the results of their educating influence on the automatic mechanism will remain though they are themselves removed. Hence the individuality of habit might still remain to some degree in the mutilated bird; and such residue of habit, individually acquired, would not in any way invalidate the view that the cerebral hemispheres are the sole control-centres, and the sole organ of consciousness. What would invalidate this view is evidence going to show that the hemisphereless animal is able to profit by experience and can be educated. If we eliminate the improvement of condition in the patient, due to the gradual recovery from the shock of the operation, such evidence does not appear to be at present forthcoming.

We may now turn to some of the more essentially psychological aspects of the scheme that has above been presented. The states of consciousness that accompany the activity of the centres V, M, and T have a two-fold aspect. In the first place, they are symbolic of occurrences in the environment which affect the organism through the afferent nerves. From this aspect we have already considered them, in what has been said concerning impressions. In the second place, they carry with them what is figuratively spoken of as a "tone" of pleasurable or painful feeling. Primitively, and in the lower organisms, control is determined by the predominance of pleasurable or painful tone in the sensory centres which are at any time conspiring to influence the centres of control. For man, in so far as he is a reflective being who frames ideals of conduct, this statement is too crude, and is contradicted by experience, unless we extend the meaning of the words "pleasurable" and "painful" in a way that can scarcely be regarded as satisfactory. This matter, however, we must also defer for future consideration. Here it will be sufficient to generalise the statement above made, and to say that the nature of control is determined by the predominant emotional tone in the sensory centres which are conspiring to influence the centre of control.

Now, is there any other kind or aspect of consciousness besides these two—(1.) that which is symbolic of occurrences which affect the afferent nerves, and (2.) that which may be described as the accompanying emotional tone? Do we, in enumerating all the elements of consciousness which fall under these two categories, exhaust to the full the contents of the curve of consciousness? Different schools of psychologists give very different answers to these questions. According to one school, in addition to the sensory elements and those of feeling, there is a third and quite distinct kind or aspect of consciousness, to which the name *volition* is given. According to the other school, there are no primary volitional elements. The phenomena attributed to volition may, they contend, be readily resolved into sensory elements accompanied by tone of feeling. It has, for example, been recently advanced, that what we term volition is sufficiently described as a mental picturing of an activity to be performed, followed by the actual experience of the carrying-out of the activity so pictured. To this a disciple of the volitional school would reply, that though the description is fairly accurate so far as it goes, it nevertheless quietly ignores, without disproving, the volitional element. He would probably indicate its inadequacy in the case of the volitional *non-performance* of a suggested activity; and would contend that, in any case, he finds in his own experience something quite different from sensory elements or elements of feeling, namely, a consciousness of control according as the activity in question is enforced or suppressed. Without presuming to decide between these schools, I may point out the fact that, if there be primary volitional elements, if, that is to say, the will be a matter of *direct* experience and not a complex product of reflection, this may, in accordance with our scheme, be due to the fact that there is a distinct type of consciousness accompanying *the functional activity of the control centres themselves.*

Merely throwing this out as a suggestion, and without offering at present any opinion as to the probabilities one way or the other, let us now proceed to correlate a little more clearly and definitely the physiology and the psychology of control. Physiologically, the activity of the control centre is at any moment determined by the impulses reaching it from the several centres in the cerebral hemispheres which have been thrown into a state of molecular disturbance by a stimulus. Psychologically, these molecular disturbances are, or are accompanied by, those sensory and emotional elements which contribute to form states of consciousness. Now let us suppose that consciousness is fully occupied with the nice control of some important bodily activity. Then, since the total content of consciousness in any given moment constitutes the empirical *me* of that moment, it is clear that this empirical *I* controls the activity in question. In automatic acts, in so far as they are accompanied by consciousness, such consciousness is a mere spectator, but in controlled activities consciousness is more than a spectator—it takes the helm and guides. And consciousness is free to guide in accordance with the laws of its inherent nature, save in so far as it is thwarted by some external constraint. In the free-will controversy, which we cannot enter into here, much confusion often arises from a failure to distinguish between the real freedom of the individual to control his activities in accordance with the laws of his inherent nature, and a supposed freedom to transcend these laws. But what we have to notice here is that, as a matter of psychology, motor control over the bodily activities is a function of, and is determined by, that complex state of consciousness which we call the empirical me.

There is another aspect of control—and one of great importance in psychology—which we must now briefly consider. We may fitly introduce it by asking the question:—How far can control be exercised over the attention? We may describe attention as the bringing of something to the focus of consciousness, and the holding it there. In the chapter on Suggestion and Association, we considered at sufficient length the manner in which impressions and ideas are brought to the focus. Now, with regard to impressions, of which we may take visual impressions as the type, there is a large amount of motor adjustment necessary in order to make and to hold an object focal: and this is undoubtedly under definite and accurate control in the sense in which we have throughout been using this word. Suppose we thus make focal a distant church spire in the midst of a visual scene. Although the details of the adjustment are carried out by the lower centres, the fixing of the eyes on just that particular object, and making it focal, is a matter of conscious cerebral control. But now, suppose that subsequently under other circumstances, sitting perhaps in our armchair, an image or idea of the church spire is brought to the focus through association. How far are we able to hold it there, and in what way is control exercised in this case? I suppose it is a matter of universal experience, that we can not retain it in the focus as idea, with anything like the same ease with which we could retain it as impression. We may, for our present purpose, neglect such facts as that not improbably, while the idea is in the focus, the control centre which was concerned in focussing it as an impression is thrown into sub-dominant activity; and that the appropriate motor adjustments may in some cases be actually made, the eyes being focussed as if for the reception of the impression, and thus appearing to an onlooker to be fixed on vacancy. These subsidiary matters we may neglect, that we may emphasize the essential question:—Can the cerebral hemispheres which exercise control over the lower automatic centres exercise control also over their own activity? In other words, Can they exercise control over their own sensory centres, and thus modify or guide the flow of ideas?

Before attempting to answer these questions, let us clearly remember the sense in which we have employed the term "control." It is motor control, control over the mechanism of automatic activity, that we have been dealing with throughout. And we have just seen that in attention, as directed towards impressions, there is this control over the motor apparatus concerned in focussing. But when we ask whether we can exercise control over the flow of our ideas, it is clear that we are no longer speaking of direct motor control. The only possible way in which we can have *motor* control over ideas, is through the regulation of the vascular supply of the sensory centres in the cerebral hemispheres. It is just possible that this may take place. It is well known that the supply of blood to the various organs and parts of organs is regulated to a nicety, for the most part through the intervention of the automatic mechanism.

It is also well known that the functional activity of any part is enhanced, within normal and natural limits, by increased blood supply. And though as a rule, in most people, the regulation of the calibre of the blood capillaries is not in the least degree subject to conscious control, it is at least physiologically conceivable that there are in the cerebral cortex control centres for the vascular supply of the sensory centres within the hemispheres, and that through them the activity of the centres may be increased, and the ideas therein rendered more vivid and lasting. There is at present no definite evidence that this is the case.

Is there, then, any control of the cerebral hemispheres over the activity of their own sensory centres? We shall do well, perhaps, to regard the matter from the psychical side, and in doing so we must bear in mind how states of consciousness are determined. Let us represent the states of consciousness in three successive moments, thus:—

1	2	3
A	C	D
b	b	c
c	*d*	*m*
&c.	&c.	&c.

Then, supposing that A b *c* &c. represents the total psychical content of the wave of consciousness in the first moment, this represents also the *me* of empirical psychology in that moment. But A b *c* &c. of moment 1 determines (subject to the qualifications given in a previous chapter, p. 82) C b *d* &c. of moment 2, which constitutes the empirical *me* of that moment. Similarly, C b *d* &c. of moment 2 determines D c *m* &c. of moment 3; and so on throughout any given series of moments of consciousness. In other words, it is the empirical *I* which constantly determines the sequence of ideas. That same psychical *I*, which takes effect through the control centres on the motor activities of the body, takes effect also through association fibres on the sensory centres which are concerned in the production of ideas.

The question, however, still remains, whether there is any power of augmentation or of inhibition of the activity of sensory centres. That is to say, supposing any given idea or image is dominant, is there any mechanism for either augmenting or damping-down the activity of the centre with the molecular disturbance of which the idea is associated? We cannot at present answer this question from the physiological point of view; and psychologically the evidence is not conclusive, one way or the other. The most we can say physiologically is, that there is no *a priori* reason why such inhibition, and its correlative augmentation, should be impossible. It is unquestionable that the inhibition of motor activities is the result of the action of the control centres, not directly on the muscular mechanism, but indirectly on the lower centres for the automatic co-ordination of motor activities. The augmentation or inhibition is exercised by nerve-centres on nerve-centres, and there is therefore no *a priori* reason for rejecting the hypothesis, that the activity of sensory centres can be augmented or inhibited by special control centres developed for that special purpose. More than this in the present state of brain-physiology we cannot say.

And, psychologically, the evidence is not conclusive. We have to distinguish between relative increase or diminution of conscious intensity through the direction or diversion of attention, and the absolute increase or diminution thereof through augmentation or inhibition. And this is by no means easy. It should be clearly noted, however, that in the absence of such control over the sensory centres as we are now considering, the direction or diversion of attention is wholly dependent on motor control over the parts by the instrumentality of which impressions are received. There can be no control over the sequence of ideas. Whether, as a matter of observation, there is any such control over the centres which are concerned in the production of sense-impressions, it is difficult to decide. Some time ago I chanced to have to wait for half-an-hour or so between two bands about equidistant from me. I amused myself by making some experiments on my power of attention, and very shortly found that I could either attend to one band or the other to the practically complete neglect of its rival, or could combine the two in an excruciating jumble of discords. With my attention well fixed on either of them, I experienced little or no discomfort from the sounds produced by the other; but when I allowed both to occupy my attention, the effect was most disquieting. I may say that in such an experiment the difficulty is, when the combined clash is in focus, to select the strains of one of the two bands; when once this is effected, the holding of those strains in focus is comparatively easy. I suppose most people who have a fair musical ear can, during the singing of a chorale, select any one of the four parts and make that focal, the other parts being accompaniment to the selected part. How is this selection effected? It is difficult to see how any motor control over the parts concerned in audition could effect a differentiation between the soprano or the bass of a chorala as dominant in consciousness.

It need scarcely be said that there is a common and wide-spread belief that attention involves something more than motor control over the parts concerned in the reception of stimuli. Such a belief, founded as it is on wide if not always very discriminating self-observation, cannot be wholly neglected. There are, moreover, not a few who believe that they can to some extent control their feelings, and their sensitiveness to pain, physical and other, and that such control is susceptible of improvement by practice. In either of these cases it is difficult to see how control can be exercised, unless there is some cerebral provision for augmentation and inhibition through control centres.

Among the commonest phenomena of the hypnotic state are suggested hyperæsthesia, on the one hand, and anæsthesia on the other. In some cases the senses become, through suggestion, extraordinarily acute; in others, they seem to be abnormally dulled and blunted. In suggested hallucinations, under hypnosis, certain centres seem to be rendered extraordinarily susceptible to what we termed internal suggestion, so that the suggested images assume the vividness and insistency of impressions. It may be that these phenomena of the hypnotic state are partly caused by some at present unexplained influence on control centres, which may have for their function the augmentation or inhibition of the activity of those cerebral centres which are concerned in sense-experience.

It is now time to sum up the conclusions we have reached, and the suggestions we have made, with regard to automatism and control. The distinction I have drawn between automatic activities and controlled action is a sharp one. On the scheme I have put forward, automatic activities involve afferent impulses, co-ordinating centres, and efferent impulses. In the performance of these activities the organism is an automaton, and the whole matter so far is purely biological. Control of the motor activities involves, and must always involve, a loop-line, in the course of which there are developed certain centres, called control-centres, whose function it is either to augment or to inhibit the lower co-ordinating centres of the automatic mechanism. Associated with these control centres of the loop-line there are sensory centres, the functional activity of which is conscious or is associated with consciousness. These sensory centres are so disposed on the loop-line as to determine the nature of the activity, augmenting or inhibitory, of the control centres according to the emotional tone associated with the functioning of the sensory centres. The sensory centres and the control centres are, in the higher vertebrates, situate in the cerebral hemispheres of the brain. So far the control is entirely motor; and so far there is much in the anatomy and physiology of the brain and central nervous system in support of the hypothesis here put forward. It is further suggested as a possibility—at present unsupported by, but so far as I know not negatived by, cerebral anatomy and physiology—that there may be cerebral centres for the control of the activity of the sensory centres. This hypothesis is put forward on purely psychological grounds. It is contended that if there is any control of cerebral activity and its mental concomitants,—if there is any faculty of attention other than that involved in the admitted motor control over the parts concerned in the reception of sense-stimuli, this must be effected by means of a loop-line, with its associated control centres. For control is due to augmenting or inhibitory impulses, and such impulses must come from outside the system so controlled. This last proposition may seem to stand in need of proof and elucidation, but I confess that it seems to me to be self-evident. For suppose a sensory centre be thrown into a state of activity, and that its functioning is associated with some form or mode of consciousness, I fail to see in what way it could be self-augmenting or self-inhibitory. We may assume that the activity is either painful or pleasurable. One can understand how its pleasurable or painful tone may differently affect a centre lying outside it with which it may communicate. But I am unable to understand how its pleasurable or painful tone could either increase or damp-down its own activity. Such a supposition appears to involve the illegitimate assumption, as I deem it, that the pleasure or the pain is something external to the activity, and can exercise control over it. Hence, for a psychology that endeavours to work hand-in-hand with physiology, the control of any functioning centre must be due to impulses, augmenting or inhibitory, coming to that centre from some other centre external to it.

Nothing has been said in this chapter concerning automatism or control in the higher invertebrates, such as the bee or the ant. Their actions seem to warrant the belief that in them, too, there is—besides the mechanism for automatic co-ordination—a mechanism of control. But at present nothing is known of definite control centres in these organisms, supposing such centres to exist. There is here a fruitful field for investigation, if we could only find a satisfactory point of departure.

In conclusion, it should be noted that control, whether it be motor-control or the suggested intra-cerebral control of the attention, should be psychologically distinguished from the sense of effort by which it is sometimes accompanied. Such a sense of effort is almost unquestionably a sensory product of the motor activities. The knit brow, slightly clenched teeth, somewhat strained respiration, and other motor concomitants, not only of motor-control but also of close and difficult thought and attention, are not without their psychical effects. And these we name effort. The part which motor impressions play in psychology, though often neglected, is of the utmost importance, as we have already seen in our discussion of Synthesis and Correlation.

CHAPTER XII.

INSTINCT AND INTELLIGENCE.

I HAVE already described some of my experiments and observations on young chicks. I propose in this chapter to call these and other similar observations to my aid in endeavouring to make clear the relation of instinct to intelligence.

The eggs from which in due course the chicks emerged, were taken from the hen two or three days before the time of hatching was fulfilled, and were placed in an incubator. The little birds, which were for the most part of good crossed breeds, with strains of Plymouth Rock, Dorking, and Game, had therefore no parental help in gaining some experience of the world. I first directed my attention to their powers of seizing and swallowing. Selecting one about eighteen hours old for definite experiment, I placed before him three small pieces of white of egg, moving them about a little in front of him with a long pin to draw his attention to them. He soon pecked at one of these, and seized it at the fifth attempt, swallowing it a little awkwardly. The next he struck at the second attempt, but not fairly, so that it was thrust aside. Transferring his attention, therefore, to the third piece, he seized it and swallowed it at the third attempt. An hour later I tried him again with egg and crumb of bread. He generally struck the morsel at the second or third peck, though he sometimes failed to seize it. Once he seized and struck at the first attempt. The observations on this chick are, I think, typical. The pecking co-ordination in young chicks is fairly accurate, but by no means perfect, at birth. They generally strike a little short. If the birds are kept for a longer period in the drawer of the incubator before they are given food,—and this may be done without injury to them, for they are hatched with a considerable store of food-yolk,—the pecking is rather more accurate, but not quite accurate.

Towards the close of their first day of active life, I caused a small fly to walk across my experimental poultry-yard in front of the chicks. Most of them took no notice; but one, whom I will call Blackie, followed and pecked at it. He caught it at the seventh attempt, and ate it; an hour later he caught another at the fourth peck, and subsequently a blue-bottle after twelve shots. This, however, he dropped and left uneaten. The others took no notice of flies, though they occasionally pecked at the disabled blue-bottle, without eating it. With a subsequent brood I tried throwing in a large fly, from which a portion of the wings had been, of course painlessly, removed. This was on their second day. One followed it, but then stopped and gave the characteristic danger note—the first time I had heard it in the brood. Perhaps the buzzing noise made by the fly called forth this note of alarm. The same chick subsequently followed the fly, caught it after several pecks, and ate it.

These experiments and observations seem, therefore, to show that the skill in seizing is not perfect at birth, and that some practice is necessary. I have spoken so far only of morsels of food; but I soon found that the chicks would peck at almost any small object I placed before them, and if small enough, almost anything was eaten, or at least tested in the bill—grain, sand, crumbs, or little bits of a chopped-up wax match. All the chicks pecked repeatedly at the letters, especially the large capitals, on the newspaper with which the experimental yard in my study was floored. I threw to my chicks about equal proportions of millet, canary seed, groats, and pari. They were about equally pecked at, and without apparent preference. The pari was too large for them, and the millet and canary were most eaten,—probably because they were smaller and more easily swallowed. Sand and fine gravel were pecked up and swallowed quite as eagerly as grain. These are, of course, necessary to the bird for triturating the food in the gizzard. Gradually a preference for groats was established, and on the fourth day this kind of grain was generally selected from among the others. With regard to other objects, they would peck at, but showed more discrimination as to swallowing, all sorts of things—pellets of paper, buttons, beads, bits of limestone, cigarette ash. They pecked at their own and their neighbours' toes repeatedly, each other's eyes occasionally, especially the eyes of the lighter-coloured birds, in which they were conspicuous; the black-beaded head of a pin, the end of a match, the point of a pen-knife, a gold seal, my ring, and so forth. All were pecked at and examined, but the larger objects with some timidity. An ordinary Bryant & May wooden match, for example, was for a time too fearsome an object for any but Blackie to tackle. On the third day four of them pecked at a burning cigarette end more than once, but sometimes were stopped by a whiff of the smoke, and then shook their heads and wiped their bills in an exceedingly comical fashion. After a minute or two they went off, but returned occasionally. When the cigarette was burnt out and cold they came and looked at it, and in one case the chick, after looking but not pecking, wiped its bill.

I did not give them water till the morning of the second day, when they were from twenty to thirty hours old. I then placed a shallow tin of water among them. Of this they took no notice. Several happened to run through it, but still took no notice. Then one chanced, as he stood in it, to peck at his toes. He at once lifted his head in the characteristic way and drank repeatedly. The others still took no notice; but presently Blackie stood at the side and pecked at a bubble near the brim, and then he drank. It seemed as if the stimulus of water in the bill at once suggested the action of drinking. As he stood and drank, others came up and pecked at the troubled water, and then they too drank. I have already mentioned how one of the chicks was subsequently caused to stop and drink at the suggestion of running through the water.

I will here transcribe a page or two from my note-book descriptive of experiments and observations on young ducklings. After piping for some time within the egg, they were hatched respectively at 1 P.M. (*a*) and 5 P.M. (*b*).

On the next day, at 12 noon, I took them out of the incubator drawer and tried them with white of egg. The co-ordination for pecking was imperfect, and when a piece was seized it was mumbled rapidly and then shaken out of the bill unswallowed. They were both unsteady upon their legs, often tilting over backwards on to their tails, but *a* was decidedly the stronger. They both pecked with uncertain aim at anything which caught the eye, such as marks on the basket in which they had been placed for warmth, grain, sand, and so forth. I placed a shallow tin of water before them. They walked through it several times without taking any notice. I dipped *a*'s beak into the water. It then drank with characteristic action, and pecking at the water drank again and again. Presently *b* imitated and drank repeatedly. Both pecked at white of egg held in forceps and seized at the third or fourth shot, shaking it out of the bill; perhaps some was swallowed. I then placed them in the warm basket.

At 2.15 P.M. I took them out. They waddled about and came to the water. Both at once drank. They pecked at white of egg, placed on a black tray to make it more conspicuous, but shook most of it out of their bills. At 4 P.M. I put *a* in a bath. He floated and kicked vigorously and excitedly, dropping excrement. In less than a minute he swam round and pecked at the sides of the bath. I then took him back to my study. Both *a* and *b* shortly afterwards came directly to the tin of water and sat in it. They pecked without suggestion at pieces of white of egg on the tray, shaking their heads and mumbling it in the bill, but swallowing freely. They pecked at grain (pari), but shook it out without swallowing it. One of them scratched his head, and tumbled over in the process two or three times.

At 6.30 P.M. they ate white of egg freely, the pecking co-ordination being much more accurate but not quite accurate. I placed *b* in the bath. He too kicked excitedly and dropped excrement; but then swam about vigorously, pecking at the sides. The co-ordination for swimming is apparently much more perfect at birth than that for walking.

At 9 A.M. next morning both *a* and *b* made at once for the water in their tin, drank, and sat in it. They ate keenly of white of egg, swallowing large morsels. Both occasionally scratched their heads, tumbling down each time. They smoothed the down of the breast after the manner in which every one has seen ducks make their toilet, applied their bills to the base of the tail, and rubbed their heads across and over their backs in the approved duck-fashion. They stood up and shook their wings, sitting down from imperfect co-ordination. At 11 A.M. I gave them a winged fly, which *a* followed and pecked at several times, but failed to seize. It then got under the newspaper, and when I routed it out *a* again followed, pecking repeatedly, till it escaped through the netting round the yard. I placed it again within the duckling's reach, and he caught it at the third peck, swallowing it apparently with much satisfaction. The other duck, *b*, caught a second fly after numerous abortive attempts. They pecked at and seemingly swallowed their own and chicks' droppings, and the signs of dislike were much less evident than with chicks. At 2 P.M. I tried them with a number of odd things, bits of paper, chopped-up matches, leaves, small flowers, small stones, red currants,—any thing of suitable size I could lay hands on. Each was seized and mumbled, and then either rejected or swallowed.

All this seems to show how necessary experience and the individual testing of things is both to young chicks and to young ducks. They have to find out for themselves the nature and value of everything they come across; but they learn rapidly and surely. They bring with them into the world an inherited aptitude for the performance of a number of activities which are essential to the maintenance of life; but they do not seem to have any inherited acquaintance with the nature of anything. Mr Spalding, who many years ago made experiments with young birds, and obtained results in some respects different from those which I have obtained, describes the instinctive terror of young turkeys when they heard the cry of a hawk. I do not question the fact that their action betrayed fear, but I am disposed to question whether they were instinctively aware that the cry was that of a hawk. In any case my own chicks gave the characteristic danger churr, a most marked and peculiar note, at any loud, strange, and unusual sound, or on sight of an alarming object. It was not a little amusing to see them, now standing and churring, and now scuttling away in terror, when I introduced to their notice a large *Carabus* beetle. Nothing, however, could be more marked than the definiteness of the effect of any very startling sound. The chicks either scatter and crouch down, remaining quite still for some seconds or even minutes, or crouch where they are and remain still. The perfect quiescence and stillness of such active and continually piping little birds is very remarkable, and clearly has a protective value. But this definite response follows on *any* disquieting sound or occurrence. If I threw a large piece of screwed-up paper among them, if I sneezed, or clapped my hands, or played a sharp chord on the violin, down they dropped; and I do not think they had any inherited acquaintance with violins. On the other hand when I took one of the chicks two days old to the cat, there was not the smallest sign of fear. The little bird took no notice of the cat. Nor did the cat take any notice of the bird, so long as I held it in my hand. When I placed it on the floor she seemed inclined to go for it; but the bird remained unaffected.

Even more remarkable was the complete indifference of my chicks to the clucking of a hen. I took two chicks ten days old in a basket to a poultry-yard. I opened the basket about two yards from a hen which was continually clucking to her brood. They took absolutely no notice of her. To test whether they were in a frightened condition, I offered them some grain on my hand. They jumped on my palm as usual, scratching it, and picking up grain. We put them with a hen in a small fowl-house. They did not seem much afraid of her, though she was terribly fussy; but the whole experience seemed to be novel to them, as indeed it was. A few days later, I took three more chicks to the poultry-yard. A hen was in the fowl-house with a young brood, to which she was repeatedly clucking—some grain having been thrown in to them. I put mine down outside—that is out of sight of her but within hearing of the clucking. They took no notice whatever of the sound, but scratched about near me.

I will further illustrate their want of innate and inherited acquaintance with the things of this world by one or two more examples. I had fed them occasionally on small worms an inch or so in length. I then took similar sized pieces of worsted wool of a rich red-brown colour, and threw them among my chicks. The avidity with which they were seized was remarkable, and most exciting were the chases after the fortunate birdling who had secured a worsted worm. I could not succeed in satisfying them with worsted, and eventually desisted lest my experiments should lead to serious indigestion. Some hours later I cut off a piece six inches long and threw it among them. Instantly there was the danger churr, and to a chick they feared to tackle that monstrous worm. Then I gave them a somewhat smaller piece, four inches long. This they regarded doubtfully; but one at length picked it up and ran off with it. There was much pulling of it one from another, but soon it was dropped. Occasionally it was picked up again and run off with, but eventually it was left unnoticed. The pleasures of eating worsted began to pall. I threw in small pieces, but they excited little interest; one was run off with and soon dropped, but eventually eaten. Two others were allowed to remain untouched. I left the four-inch piece. Presently I was roused from my writing by sounds of excitement and little pattering feet. Blackie had seized the piece, and was being chased for the prize. Escaping from the "yard," in which the chicks were confined, by leaping over the fender, he ran to the corner of my study, and after some struggles with the worsted succeeded in swallowing all the four inches.

It may be said that to supply worm-like worsted was a piece of base and unnatural deception. I will give one more case in which there was no such deception. I was desirous of ascertaining whether my chicks had any innate awareness of the difference between a large fly and a hive-bee. Now Blackie was intimately acquainted with flies, large and small, and liked them well. When I placed a hive-bee in my experimental poultry-yard most of the chicks were afraid of it, as indeed they were of large flies; but Blackie without hesitation snapped it up and ran off with it; then he dropped it, shook his head much, and wiped his bill repeatedly. As I mentioned before, I do not think he had been stung; if so, he quickly got over any ill effects, and was happy and eager about other things in a few minutes; more probably he had tasted the poison. In any case he no longer took any interest in that bee. Some hours later on the same day (his fifth) I placed beneath a glass tumbler a large fly and a small humble-bee, both winged. Blackie and another chick pecked at both, seen through the glass. I then let the bee escape; Blackie snapped it up, ran off with it and soon swallowed it. Another small humble-bee he seized at once, disabled it by dashing it against the ground with his bill, and swallowed it. Both of these humble-bees had stings. I threw a humble-bee among another brood of chicks ten days old. They pecked at it with rapid strokes, jerking it on one side. At this stage of development they approach anything new and strange to them with some suspicion, and peck at it in this way. For example I gave them some stewed currants and raspberries; these were struck at rapidly, and thrown aside. Presently one was picked up and run off with, other chicks giving chase, and eaten. A few currants were subsequently eaten, but with a good deal of shaking of the head, and apparently not much relish. The ducks ate them rather more freely. One of the ducks picked up the disabled humble-bee, and after mumbling it for some time in the water swallowed it.

I could generally in their early days induce my chicks to take anything, and even to seize again something they had found distasteful, by tapping with my finger or a pencil on the ground to direct their attention to it, in imitation of the pecking of a hen. In this way, I induced one of my chicks to take up the beelike *Eristalis* which it had so far avoided. Directly he had seized it, others gave chase, and one took it from him and swallowed it. Next day I threw in among them a hive-bee and an eristalis. Both were left untouched. I drew their attention now to one, and now to the other, by tapping near, and moving it with my pencil;

but without effect. I may say that tapping was ceasing to be effectual, as I had in this way frequently indicated things which they had not always found pleasant.

I have generally dismissed my chicks to the poultry-yard at the age of ten days, and in no case have kept them beyond three weeks. But I have never seen any chick, nor any duckling, catch a fly on the wing. There are often a good many flies hovering over, and settling in the experimental yard. I have very frequently seen the chicks, and occasionally the ducklings, strike at the flies in the air, but never successfully. I have experimented by placing flies in a tumbler covered with a card. The chicks then peck repeatedly at the flies. On removing the card the flies escaped, and the chicks have pecked at them as they flew, but have not caught them. I also inverted a tumbler over flies, and removed the tumbler when they were on the ground. Blackie on one occasion caught one before it flew; but neither he nor any other of my chicks has been successful in catching a fly on the wing.

I have now described, perhaps in undue detail, a few of my observations as noted down at the time. To some they may seem trivial, and scarcely worth the making and the noting. To us, as students of comparative psychology, their interest lies in the light they throw on the beginnings of psychical life and activity in the chick or duck. Let us now see how the observations serve to bring out the relation of instinct to intelligence. But first, What is instinct? How shall we define or describe it? Let us go back to the little chick in its early efforts at pecking. Here we have a motor response to a certain stimulus. And there can be very little question that the motor response is, as we are apt to say, purely mechanical,—or, as we should more correctly say, purely organic. It is a bit of animal automatism, not necessarily involving more than the lower brain-centres. But it is a bit of automatism which is accompanied by consciousness in the broader sense in which I have used this word. And the *rôle* of consciousness in the matter of pecking is to select the adequate responses, and to steady the muscular mechanism to its work. Let us describe the automatic activity as due to *innate capacity for motor response.* Then in the animal kingdom, we find that the automatic responses which are the outcome of this innate capacity are variable in their adequacy. The chicks at first made bad shots as well as occasional good shots. Now the greater the variability, and the greater the initial percentage of inadequacy, the more necessity is there for the acquisition of skill by the individual. On the other hand, the less the variability, and the smaller the initial percentage of inadequate response, the less the necessity for any individual acquisition of skill.

And now we can give a good working definition of instinct in its objective aspect, and at the same time indicate the relation of intelligence to instinct. Instinctive activities, in their theoretical perfection, are those in which there is no variability, in which the percentage of inadequacy is *nil*, and in which, therefore, there is no necessity for the individual acquisition of skill through the control and guidance of intelligence. If the chicks had pecked perfectly from the first they would have had this instinct in perfection. As it was, they required a little intelligence, acting by and through experience, to perfect their activities. The instincts were very nearly, but not quite, perfect.

A qualification may here be introduced. It is possible that, in the absence of intelligent control, the repetition of the activity would render it more accurate. In other words, the inaccuracy may have been partly due to a want of *organic* smoothness of working in the automatic mechanism. There can be no question that many activities, which are in due time performed in virtue of the automatic mechanism for response, are not and cannot be performed at birth, because the organic development is not then sufficiently advanced. The flight of birds is, according to Mr Spalding's observations, instinctive. He kept young swallows caged until they were fully fledged, and then allowed them to escape. They flew straight off at the first attempt. They exhibited the instinctive power of flight in a practically perfect form. But it is clear that such a proceeding would have been absolutely impossible for the callow young when they were first hatched. The instinctive power is in these and like cases deferred until such time as the organic development renders the automatic performance of the activities physically and physiologically possible. The automatic activities, which are the outcome of completed sexual development, are good examples of such deferred instincts.

It is a necessary corollary of the view here advanced that in instinct *as such* consciousness is a mere epiphenomenon —a by-product, with no bearing whatever on the performance of the activity in so far as it is instinctive. An organism—if such exists—in which all the activities are throughout life purely automatic and purely instinctive, might indeed be conscious, but its consciousness would be of no practical value; for all the activities being, *ex hypothesi*, automatic, there would be no conscious guidance or control. Consciousness might be present as a spectator of the activities; but it would be a mere spectator without power of guidance, since, in so far as guided by intelligence, activities cease to be instinctive. It should be clearly grasped that, in so far as an activity is guided by individual control towards more complete accuracy, just so far does it cease to be instinctive, as the word is here used, and become intelligent. And when an instinct is, as so often is the case, modified and adapted to meet new circumstances, the modification and adaptation is no part of the instinct as such, but is due to intelligent control.

I repeat, then, that in instinct as such consciousness is an epiphenomenon or adjunct. But this does not, of course, imply that it is absent. Only in so far as consciousness accompanies the performance of instinctive activities can intelligence get a hold on them for the purpose of control and guidance. The performance of automatic activities affords to consciousness data, which form a foundation upon which the psychical structure reared by intelligence is based.

So far as vertebrates are concerned it is only in the lower forms of life, and in those which are born in a relatively advanced state of development, that we find the products of what I have termed the innate capacity for motor response in their relatively perfect and, so to speak, crystallized form as instincts. When little alligators just hatched, and still adhering to the shells by their umbilical cords, briskly show fight when approached, dragging their shell behind them and rushing with open jaws at any thing presented to them and madly biting it;* or when little snakes taken prematurely from the mother's oviduct threaten to strike, and make the characteristic burring noise with the tail,†—these are instances of unquestionable instinct. In young chicks the following responses are from the first sufficiently well-defined to be regarded as instinctive. Pecking, walking, scratching themselves, preening their down and feathers, stretching up and clapping their wings, scratching on the ground, squatting down and "dusting" themselves, scattering and crouching when alarmed, making the danger churr and other sounds previously described. In addition to these there are numerous minor activities which are distinctly chick-like, and differentiate the chicks from young ducklings. In the ducks, the way they seize and mumble the food in the bill, their swimming, their piping, their mode of smooth-

* As observed by Mr Devenish. See *Nature*, v. 47, p. 587.

† As observed by Mr W. Larden in the "Vivora de la Cruz." See *Nature*, v. 42, p. 115.

ing the down of the breast with their bills, applying the bill to the oil-gland, rubbing their heads across their backs, and other minor characters, are definite enough to be termed instinctive. But when we come to higher animals, especially those born in an immature condition and dependent at first upon parental care, it is not so easy to say what activities are definitely instinctive. The basis in innate capacity for motor response is unquestionable; but imitation and individual guidance, through intelligent adaptation, are factors which render the activities less purely instinctive. I am inclined to regard imitation and tradition, especially in animals which live in flocks, packs, or herds, as of very great importance. By tradition I mean this: that the animal is born into a group of animals which perform a number of activities in certain ways, and that through the imitative tendency it falls into these ways, which are thus handed on or carried down through tradition. In a very interesting chapter on "Fear in birds," Mr W. H. Hudson, in his work on "The Naturalist in La Plata," describes many observations which point to the conclusion that the fear of man and other enemies is very largely due to parental instruction or to individual experience. And my own observations tend in the same direction. Mr Hudson uses the term "tradition" in the sense I have above adopted.

In any case I am quite clear that we must not be too ready to put down to instinct all the habits of animals, even definite habits common to a species, without taking the trouble to ascertain by careful observation and experiment how far these habits, though based on an innate capacity for motor response, are rendered definite through imitation, parental teaching, or tradition.* I believe that there is here

* The complex problem of the migration of birds is a case in point. It is unquestionably one that involves more than instinct, though it has an instinctive basis. What is the nature of the internal craving which prompts to migration probably no one but the bird knows—and the

an extensive field open to *experimental* observation. Is it too much to hope that in connection with some university in Great Britain there will some day be founded a Chair of Comparative Psychology, with an experimental station attached, where animals may be reared under test conditions and under foster-parents, so that the development of their intelligence may be carefully studied?

No doubt in those species of which the normal environment is relatively simple, and but little subject to variation, the responses due to inheritance will be relatively definite and instinctive in their nature. With increased complexity and variability of environment will come decreased definition in the inherited modes of response, such definition being left to individual acquisition. In man, therefore, with his exceedingly complex and variable environment, the number of definitely instinctive activities is few; while at the same time the innate capacity for motor response forms a broad and ample, if relatively undifferentiated, foundation for an indefinite number of activities, which owe their guidance to the fostering care of parents and nurses.

Omitting now, for the present, any further reference to imitation, tradition, or parental guidance, and fixing our attention on the relation of instinct to intelligence, we may say that animals and men alike come into the world with an innate capacity for active response to certain stimuli. This

---

bird only feels, but knows not. Nor do we know what the stimuli are which start the motor activities. In the far simpler case of the migration of recently-hatched eels (elvers) up the streams from the estuaries, the continued stimulus of flowing water may be the determinant of the continued upward and onward progress. I have before suggested that highly developed and differentiated sensations, through the constant use (by the individual) of the membranous labyrinth and semicircular canals, may be a factor in the direction of migration. But in any case there is probably a traditional element of great importance, the tradition having perhaps continued unbroken from a time when the configuration of land and sea differed from that which now obtains.

is part of their organic inheritance. The response may be from the first an accurate and adequate response: in such cases we term it instinctive. But more frequently the responses have a variable amount of inaccuracy and inadequacy; in such cases the animal, as a matter of observed fact, has a power of selective control over the responses; and this power of selective control over the activities which are essential to daily life, is the first stage of intelligence. And whereas for the instinctive action, as such, consciousness is only an epiphenomenon or adjunct accompanying the performance of the action, for its intelligent guidance consciousness is essential.

When the chick, so soon as it is taken out of the drawer of the incubator, strikes with fair but not complete accuracy at a piece of white of egg, there can be little question that this action—complex as it is, and involving as it does the due co-ordination of a great number of muscular impulses—is of the nature of an inherited organic response to stimulus; that the sight of the small white morsel is just the touch of the trigger that, so to speak, fires off this complicated train of activities, the ability to perform which is innate. But we have every reason to suppose that the performance of these actions is accompanied by feeling or consciousness. So that in these early days of life the consciousness of the chick is, if one may say so, entering into and taking possession of its organic inheritance. And consciousness, like a wise heir, at once proceeds to set its estate in order and to remedy such imperfections as it finds therein. In the case of the chick, the inheritance is already so well organized that it requires very little individual control of consciousness to put things in excellent working trim. In the case of the human infant, however, there are noteworthy differences. In the first place, the heir comes into possession when he is, as compared with the chick, far younger and less mature; in the second place, his inheritance is of vastly greater extent, with commercial relations of far greater range and complexity; and, in the third place, it has been the custom for generations of his ancestors that during his minority he should be aided in the administration of his estate by faithful stewards, and should be instructed therein by wise tutors. Still, notwithstanding these differences, it remains true that the infant consciousness, like the chick consciousness, has—more gradually, no doubt, and with more external aid—to enter into and take possession of its organic inheritance; and, no matter how much he is aided and instructed, he has to do so individually and for himself. None can share this task with him, or perform it for him. With this inheritance, moreover, he must make the best of life. No kindly uncle or aunt can bequeath him a new estate. The inheritance is his to deal with as he may and can, within the assigned limits; for his very power of dealing with the inheritance is part of the inheritance itself.

We are getting, however, too far from our foundations, and must return to the stage up to which we have, I trust, securely built. The chick, or the child, in the early hours or days of life acquires skill in the management of that part of the organic inheritance which we call its bodily organs. And this skill, rapidly in the chick, slowly in the child, reaches a high pitch of delicacy and exactitude. The chick of a week old will pick a fly off your finger, and not so much as touch the finger itself. The method employed is that of trial and error. What we term the control over our activities is gained in and through the conscious reinforcement of those modes of response which are successful, and the inhibition of those modes of response which are unsuccessful. The successful response is repeated because of the satisfaction it gives; the unsuccessful response fails to give satisfaction, and is not repeated. Sufficient has, however, been said in the chapter on the sense-experience of animals concerning the *rôle* of consciousness in the acquisition of skill. It is a vitally important matter in the psychology of animals, and involves the selective activity of intelligence. But we need not here repeat what has been already discussed on previous pages.

Let us pass, therefore, to another aspect of intelligent guidance, and in order to avoid referring back to instances before given in this chapter, let me further describe some observations on the young chicks, mention of which was made in a previous chapter. On the evening of their third day of active life, I placed before my little birds two objects new to their experience, a small worm and a yellow and black caterpillar—that of the cinnabar moth, so common in the summer on ragwort. The birds looked a little timidly and suspiciously at both of them. So far as I could judge, they were not more suspicious of one than of the other; they were probably suspicious of both, because the objects were rather larger than those which the chicks had been accustomed to peck at, and because they moved. They pecked at them timidly once or twice; but as it was getting late, and my chicks were rather sleepy, I felt it my duty, as their acting foster-mother, to put them to bed. Next morning, when they were fresh and vigorous, I repeated the experiment. Again both objects, the worm and the cinnabar caterpillar, were pecked at timidly, and eventually taken up in the bill and run off with. But the caterpillar, which is known to be distasteful to most birds, was dropped at once; while the worm was, after some comical efforts, bolted. Subsequently the caterpillar was occasionally pecked at, and more frequently merely looked at; but soon it was left unnoticed. Fresh small worms, on the other hand, were at once and with confidence snapped up and carried off, causing a most exciting chase, the fortunate possessor being allowed no peace for the delightful efforts necessary for swallowing the worm. I have, with other chicks, tried similar experiments with cinnabar caterpillars and loopers, or other edible caterpillars of about the same size, and with similar results. And with ducklings the results are again similar. In no case, I may mention, was the caterpillar injured or the skin broken.

In these experiments and observations the points to be noticed are, first, the absence of any instinctive acquaintance with the difference between a nice worm and a nasty caterpillar; secondly, rapid profiting by experience after a few practical trials; thirdly, arising out of this, the discriminating by sight between the two objects; fourthly, the association of a nasty taste, or perhaps a disagreeable odour, with one of the objects, and pleasant gustatory results with the other; and, fifthly, guidance of subsequent action in accordance with the results of experience. In the last two points we have, in an elementary form, the basis of intelligent adaptation to circumstances. Intelligence, as I use the word in this work, is founded on experience; it involves the association of impressions and ideas, and it implies a power of control over the motor responses.

Let us now take a case illustrative of a rather more advanced stage of intelligence. I kept the chicks in my study, near a small gas-stove, so that I might regulate the temperature. For my first brood I made a sort of yard, paved with newspaper, and with newspaper walls propped against the fender, rugs, and so forth. (I now use wire netting.) At one side the turned-up newspaper rested against a chair. Blackie was a week old, and particularly bright and fresh, perhaps in consequence of his hearty meal of worsted. He was standing near the edge of the yard, pecking vigorously and persistently at something, which I discovered to be the number of the page of the newspaper. He then transferred his attention and his efforts to the somewhat turned-in corner of the newspaper, which was just within his reach. Seizing this, he pulled at it, bending the newspaper down, and thus making a breach in the wall of my yard. Through this breach he stepped out into the wider world of my study. I put the paper back as before, caught the errant Blackie, and placed him in the yard, near the scene of his former efforts. He again pecked at the corner of the paper, again pulled it down, and again escaped. I then put him back as far as possible from this weak place in my poultry-yard. Presently, after some three or four minutes, he sauntered round to the corner, repeated his previous procedure, and again made his escape.

Unquestionably this is a more complex case of intelligence than the one I gave before. But it is of the same order. It was founded on experience; it involved the association of impressions and ideas; and it implied a power of profiting by the experience through the association. The chick found that a certain action, performed in the first instance, it would seem, *without any view to any particular result*, produced certain effects; those effects were pleasurable; an association was formed between the idea of pecking at that corner and the idea of walking out into the room. And, subsequently, the action of pulling down the newspaper was repeated for precisely the same reason that the action of picking up the worm was repeated—namely, because it had become associated through experience with pleasurable consequences.

Enough has now been said to bring out the nature of that innate capacity for motor response due to organic inheritance, which forms the basis of all animal activities; and also the nature of those specialized forms of response, wherein consciousness is at most an epiphenomenon or associated adjunct, which are termed instinctive; and enough to show, by one or two simple examples, how intelligence guides the activities to individually adaptive results, through the association of ideas and the control of motor responses. We shall have to consider in future chapters other and more complex cases of intelligent guidance.

CHAPTER XIII.

THE PERCEPTION OF RELATIONS.

THE complexity of consciousness is so intricate, and yet so orderly, that it is difficult to trace all the varied threads of our mental tapestry. In previous chapters we have endeavoured, in our analysis, to pick out some of these threads. We have seen that in our visual impressions, sight being for us the dominant sense, retinal sensations coalesce or combine with motor sensations, and that seemingly simple impressions are therefore the products of a subtle synthesis. When we see a violet, the retinal sensations somewhat differently grouped in the two eyes, enter into synthetic union with certain motor sensations, the concomitants of movements of and within the eyeballs, whence arises an impression of a coloured and shaded object, possessed of a certain form, and located just there in the midst of a marginal field. But this impression forms the nucleus around which representative elements cluster. The visual field, how it is lighted and how the shadows fall, the relative size of the subsidiary objects, their grouping around the focal object,—all these involve representative elements which serve, so to speak, to give further body to the mental picture, and which result from a vast amount of visual experience, whereby retinal and motor elements have entered into a close co-ordination. The violet is a visual impression; but it is something more,—it is a centre of aggregation. Its fragrant odour, for example, is suggested; and for me, as I have before mentioned, a particular spot in a quaint old garden, often visited in childhood, has a tendency to come to mind. That is to say, violets have a strong tendency, through association, to form the focal starting-point of a representative visual scene.

So far we have been dealing with the presentative elements of impressions, and with their representative revivals. Impressions as such are constituted solely by presentative elements synthetically combined, though they may rapidly accrete around themselves in further synthesis representative elements; and these representative elements, so far as we have already considered them, are merely the revivals of those sensations which constitute the presentative elements. And here it may be well to add a reminder that we defined a sensation as an undecomposable element in consciousness, due to an afferent impulse, or a combination of afferent impulses. Our analysis disclosed sensations of the special senses (retinal, auditory, olfactory, and so forth), sensations of pain and of general sensibility, motor sensations, and, it would seem, sensations of distance and solidity. These latter we found to be due to the combination of afferent impulses, which are probably physiological and infra-conscious. And it will be remembered that, in reply to the question, *why* the combination of certain infra-conscious impulses gives rise to a sensation having the quality of distance or of solidity, I said that I did not pretend to know. It is like the question why a combination of carbon and sulphur in certain proportions gives something so different from either constituent as the colourless liquid carbon disulphide. We can only say that, so far as we know, this is the way things are constituted. We here reach the limits of scientific analysis. In like manner of distance and solidity, we can only say that, under given conditions of stimulation, that is the form which impressions assume. Science cannot say why. Science has to be content with the fact that, so far as it is at present able to ascertain, this is the way in which such impressions are constituted.

But the important point for us to notice in the present connection is that outness, position in the visual field, and solidity, are already given in the visual impression, and that they acquire their significance for practical conduct through their correlation with tactual, auditory, and other impressions. This correlation is based on association; and what I have just termed "significance" is the suggestion of tactual, auditory, or other representative states on the occurrence of visual impressions. In and through this correlation the presentative elements, which our analysis has disclosed, together with their representative revivals, enter into fuller and more complex synthesis; but in the synthesis, these sensation elements (and their revivals), and these alone are present. If it be said that in addition to these elements there is the synthetic consentience itself which combines them into new mental compounds, I would reply that this is not only true, but is involved in the conception of psychology as a science. We study psychology in order that we may ascertain the laws of the combinations which lead to the genesis of mental products, just as we study chemistry that we may ascertain the laws of the combinations which lead to chemical products. If there were no synthetic activity in either case, there would be no science of psychology or of chemistry. Such activities in nature are what we endeavour to give a generalised statement of in the primary laws of nature. These laws are, we believe, universal laws. The activities in question are not peculiar to and resident in individual units, whether these individuals be organisms or crystalline products of chemical combination. They are general or, in philosophic phrase, universal, and are manifested in the individual only in so far as the individual is a constituent unit in the universal. This is now freely admitted in chemistry and in biology; until it is as freely admitted for mental products, psychology will either fall below or soar above the level of science. From the strictly empirical point of view, it may be said that there is nothing in the impression or its idea but a synthesis of conscious sensations and infra-conscious impulses (or their revivals). And if a disciple of Leibnitz would bid us add, "Yes, nothing, *save the synthetic principle which combines them*," this may be freely admitted, on the distinct understanding that the synthetic "principle" is strictly analogous to the synthetic "principle" by which carbon and sulphur in due proportions combine to form carbon disulphide, or by which carbonate of lime crystallizes as calcite. And by strictly analogous I mean that in each case we are dealing with natural law. But if it be meant that the "principle" of synthesis is something supernatural,—not the inner essence of the natural process, but something superadded thereto,—then we must reject the hypothesis as unnecessary, and not in line with the results of investigation in other departments of knowledge. And this, I need hardly add, were it not that it is so easy for students of mental science to misunderstand each other,—this, I say, is no denial of mental individuality. Such denial would be on the face of it absurd; it is merely the placing of mental individuality in line with organic individuality, as the natural product of the complex interaction of natural laws.

We have now to pass on to the consideration of other factors of mental products by the synthetic incorporation of which these products take on new and higher characters.

The visual impressions are different according as the object is near or far, and located here or there in the visual field. We *explain* this by saying that the object is seen *in relation to* other objects in space. But such explanation involves the perception of the relation as such. Until the relation shall have been perceived, it is impossible to view the objects in the light of the relationship. And so long as we are dealing simply with impressions in naive sense-experience, the relationship need not yet have been perceived. The relations may have been implicitly there, but they may not yet have become explicit. They may not yet have been brought to the focus of consciousness. I look around me in my room, and fix my eyes, and my attention on this, on that, and on the other familiar object. With each new centre of vision and of attention there is a new grouping of the visual scene in relation to a new focus. But as I do so, my mind need never for a moment dwell on the relationships of the objects to each other in space. Related in space they are, and I *can* perceive the relations. But I may look around me for hours without paying any attention to them—without their ever coming to the focus of my consciousness. No doubt, the fact that such relations have been clearly grasped lends a subconscious relational tone to my mental synthesis. But for the little child, this is probably not so. For the child has not yet risen to the perception of relations; the relationships for him have, I conceive, never become focal. For him the mental products are impressions, and nothing more; for me they are percepts in so far as they carry with them this relational tone. A percept * is an impression to which is added a conscious or subconscious perception of relation to the subject or to other objects. And we have now to endeavour to ascertain the psychological nature of the perception of relations.

As I look around me, and transfer my attention from one object to another, there is at each transference a transition in consciousness. If we symbolize by letters the succeed-

---

* The term "percept," like so many other terms in psychology, is differently defined by different authors. The reader is, therefore, requested to note the sense in which it is employed in this book. Those who have read my *Animal Life and Intelligence*, will observe that I use the term here in a somewhat different sense from that in which I employed it in that work. I need hardly add that I think my present usage more satisfactory than my previous usage.

ing states of consciousness, we have a series which may be thus formulated :—

r — s — n — y — t — v.

But consciousness is continuous. The states *r s n*, &c., are not isolated from one another, but form phases in a series, one of whose attributes is continuity. And we sense or are aware of the transitions as they pass. We do not fix our attention upon them. Our attention is engaged by the substantive states themselves. Our attention passes from *r* to *s*, and from *s* to *n*, and so on. The consciousness of transition is altogether marginal, and in the case of the little child has, I conceive, never become other than marginal. Let us try and think ourselves into the condition of early childhood. There are placed before us three wafers or counters, two red and one blue. We look at them in succession, so that first a red wafer, and then another red wafer, and then a blue wafer, occupies the focus of vision and of consciousness, the sequence therefore being red wafer—red wafer—blue wafer. As we do so we feel that there is a transition in consciousness, first from red to red, and then from red to blue. When the transition is from red to red, we sense or are aware that the focal impressions are "similar;" when the transition is from red to blue, we sense the focal impressions as dissimilar. Now here the point to be noticed is that the similarity or dissimilarity is not in the focal states themselves, but in the transitions from one to another. In the absence of the red wafer, there would be no sense of dissimilarity produced by the impression of the blue one. It is only when the attention is transferred from red to red, or from red to blue, that we are aware of the similarity or dissimilarity between them. It must, however, be clearly understood that it is with the mere awareness of that which is interpreted by perception as similarity or dissimilarity that I am so far dealing. Our three wafers are moreover to be regarded as illustrative and representative of any kind of similarity or dissimilarity. Let us suppose that for the red wafers, there are substituted two sweets, and for the blue wafer, a piece of coal or chalk. A child who has as yet no clear perception of similarity may be aware of the likeness of one sweet to the other, and practically act upon this awareness, and may likewise be aware of the dissimilarity of the piece of coal or chalk to the sweets. So too a child may be subconsciously sensible of the similarity of two leaden bullets the one to the other, and of the dissimilarity of a clay marble to either of them, and yet have no clear perception of the relation of likeness or unlikeness.

It will be noticed that we are compelled to explain in terms of relationships (because we cannot *explain* in any other terms), the mental condition prior to the perception of relations. We are forced to use the words "similarity" and "dissimilarity," "likeness" and "unlikeness," for that mere awareness of a transition in consciousness, which in the light of perception will become (but is not yet) similarity or dissimilarity. We are inevitably driven to employ the language of anticipation. And I know not how better to make clear the distinction I would draw, than by laying stress upon the verb, and by saying that a little child endowed only with sense-experience, *senses* or *is aware of* the similarity between two sweets, and the dissimilarity between a sweet and a piece of chalk, but does not perceive the similarity or dissimilarity; or, to put it in another way, that he senses the sweets as similar, and the chalk as dissimilar, but does not perceive the similarity or dissimilarity as a relation. And that because the sweets or the chalk monopolise the focus of consciousness, because the child has not yet had occasion to look within on the workings of his own mind, because the faculty of perception is not yet developed.

It will possibly be said by some of those who have not quite caught my meaning, that to account for the origin of the perception of relations, I am, by a piece of badly concealed thimble-rigging, introducing into sense-experience that of which I feign to be in search. I first assert that in sense-experience, as such, there is no perception of relations; I pretend to inquire how these relations as objects of perception have arisen; and thereupon proceed to assume that there is in sense-experience a dim "awareness," or "sensing" of those transitions in consciousness which will be eventually perceived as relations; thus slipping the relational pea beneath the thimble of sense-experience. I do not think that this would be a just criticism. In any case, I am desirous of laying bare all my assumptions. I assume then that in sense-experience there are impressions and ideas; and that, since sense-experience is continuous, there is also a consciousness of the transitions between these impressions or ideas. Two objects, A and B, are placed before a being in the stage of sense-experience. The eyes are fixed first on one and then on the other. There is an impression of A followed by an impression of B. And these successively occupy the focus of consciousness. But there is also the transition from one to the other, which since it does not in any way interest sense-experience, remains marginal. And this transition it is which, when focussed through reflection, becomes the relation between A and B, the focussing of it being termed the perceiving of the relation. *The transition does not become a relation, until it is thus perceived or focussed.* The distinction is not an easy one to grasp, but will I trust become clearer as we proceed.

Similarity and dissimilarity are the most generalised relations that we know; for objects may resemble or differ from each other in a great variety of respects. If, then, these relationships are the expression of the transition in consciousness from one focal state to another, it would seem likely that all relations between impressions and their ideas are primitively given in consciousness as the transitions from one focal state to another. Such, I believe, is invariably the case. And the transitions in the *naïve* everyday life of practical experience are persistently marginal. In endeavouring to interpret therefore the consciousness of the little child, we may say that *the subconscious awareness of relations prior to the advent of perception has its origin in the transitions of consciousness from one focal state to another;* and further, that *in the unreflective flow of consciousness the transitions are entirely marginal.*

Let us now return to our symbolic representation of the sequence of focal states of consciousness, consisting of impressions or their ideas :—

r — s — n — y — t — v.

It is clear that though we may be aware of the transition, say from *r* to *s*, as it occurs, we cannot in the act of passage sense it in its completeness; for in the act of passage the transition is *not yet completed.* And when it is completed, and *s* is reached, it is the impression or idea *s*, and not the transition that led to it, which occupies the focus of consciousness. The matter may be put concisely thus :—Every relation involves two related terms; until the second term is given the relation is incomplete; but at the moment of passage from the first to the second, the latter is not yet given. Hence it is impossible to sense a relation in its completeness during the transition which is its psychological equivalent.

We have now to consider how we do come to perceive relations in their completeness, and how we obtain definite and clear-cut perceptions of relation.

Since it is impossible to get a complete and precise perception of the relation of *r* to *s* as the wave of consciousness is passing from the one to the other, and since it is necessary for the perception of a relation that *both* related terms should be given, it is clear that it is only by looking back on *the past course* of the psychical wave that we can definitely perceive the relationship which was previously only sensed. In other words, the perception of a relation (that is to say, the making of the relation focal) involves introspection, which, as we have already seen, is also retrospection or reflection. If therefore our symbolic illustration

$$r - s - n - y - t,\ \&c.,$$

be taken in illustration of the flow of impressions or ideas in sense-experience, the connecting links being marginal in consciousness, the formula *

$$r - \overparen{s - n} - y,\ \&c.,$$

may be taken as roughly illustrating the retrospection or reflection necessary for the perception of a relation, or the focussing, as idea, of a transition in consciousness.

It will be well to consider first the perception of a simple space-relation. Let us direct our attention to the relation of the point marked *a* to that marked *c*.

Fig. 13.

As we do so, our eyes travel to and fro between the two. First, *a* is focal, and *c* marginal; then *c* is focal, and *a* mar-

* The curve indicates that the idea *n* deals with the transition between *r* and *s*. There is, of course, no backward flow of the wave of consciousness, which passes onwards through r — s — n — y, &c., in orderly sequence.

ginal; and finally, the eye is perhaps fixed on the centre of the vacant space between them, both *a* and *c* being marginal. With the eye thus fixed, the visual impression is that of a pair of points; but if consciousness is focussed on their relation to each other, we have a perception of this relation as such.. But since the perception is thus directly produced by making the relation focal, and the related terms marginal, where is the justification for saying that the perception of relations involves reflection and retrospection? Remember how the perception of the relation is reached. First *a* is focussed with *c* marginal, and then *c* is focussed with *a* marginal. Then, and then only, when each related term has in turn been focal to consciousness, can we make both marginal, and focus the relation itself. In our diagrammatic scheme r — s — n, *r* is the impression of *a*, *s* the impression of *c*, and *n* the transition from the former impression to the latter; and therefore *n*, as a present and focal state of consciousness, is symbolic of experience gained in preceding moments of consciousness. Moreover,—and this perhaps is the most important feature of the process,—that which is thus symbolised is not any previous impression, but the transition between impressions. And it is difficult to see how completed transitions in consciousness could be perceived and symbolised without reflection; that is to say, without looking back upon the past course of the psychical wave. No doubt the reflection here is only embryonic and germinal. No doubt the process is only incipiently introspective and retrospective. But therein lies its chief interest; for we seem to have here, in the perception of a simple space relation, the first step into a new realm of conscious experience, — a realm of well-nigh boundless extent.

Let us now extend our spatial diagram as under:—

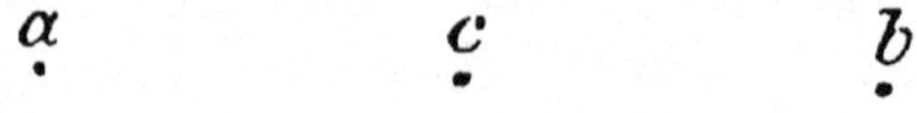

Fig. 14.

Having perceived the relation of *a* to *c*, we transfer our attention to *c* and *b*, and proceed to make the relation between them focal just as we did for *a* and *c*. Having done so we may proceed a step farther and deal with the group *a c b*. We then perceive that the relation of *a* to *c* is in all respects similar to the relation of *c* to *b*. We are here, be it noticed, *perceiving a relation between relations* and one which involves similarity of distance and similarity of direction. The case is more complex, but it does not involve any new order of perception. We are still dealing with a transition in consciousness from one focal state to another, and it makes no difference in principle whether the focal states between which the transition takes place are impressions such as *a* and *c*, or perceptions of relation such as *a c* and *c b*. In this case in our diagrammatic scheme *r* — *s* — *n*, *r* is the relation of *a* to *c*, *s* the relation of *c* to *b*, and *n* the relation of *r* to *s*.

We may now proceed a step farther, and make our figure more complex:—

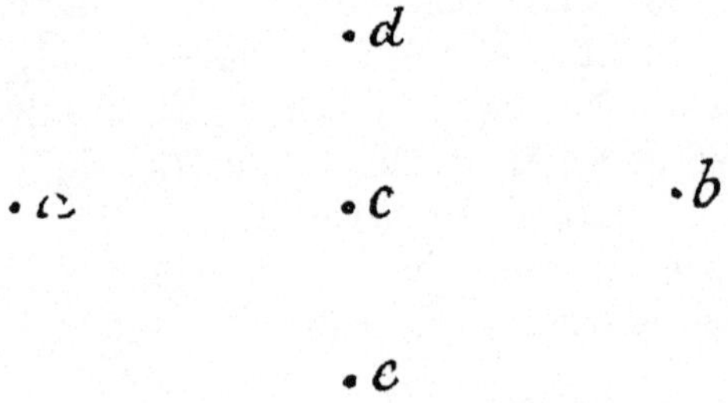

Fig. 15.

Let us direct our attention to the relations of *d* to *c* and *c* to *e*. We soon perceive that the relation of *d* to *c* is similar to that between *c* and *e*. We repeat, in fact, with respect to the group *d c e* the experience we gained with the group *a c b*. But if we take the relation of *c a* to *c d* in the group *a c d*, we find a dissimilarity in the relations, and that in two respects. For first the distance *c d* is less than the distance *c a*; and secondly, the direction *c d* is different from the direction *c a*. The case is again somewhat more complex, but it introduces no new order of perceptions, though it does show that the relations may be definitely perceived to be either (1) similar or (2) dissimilar to each other.

It should be clearly noted that in the perception of relations consciousness is merely dealing in a new way with old materials. Long before we pay attention to, and definitely perceive space relations, we are marginally subconscious of them. We are aware of them, and practically act upon our awareness, but we have not learnt the trick of making them focal in consciousness. Nor have we in our early days much need to do so, since marginal awareness amply suffices for all the purposes of practical experience. It is only when we seek to *describe* things for our neighbours' behoof or for our own satisfaction, that relations are dragged into the focus. Ask a boy of five or six to describe to you how he performs some simple and familiar operation. He will probably want to *show* you how it is done. But if you meet him with a request to be told, not shown, how it is done, you will probably soon see how puzzled he is. He cannot describe it except in terms of relationships which he has only sensed and not yet perceived. It is not the mere words that he lacks; it is the ideas which lie in an unfamiliar field. He has to look back upon his past experience to see how he did it, and he is not much given to such reflection.

Another point that should be noted is that in the case of

space relations we have a definite physiological basis for the feeling of the transition from *a* to *c* or from *c* to *d*; for the transition in consciousness, when we are dealing with these impressions, is accompanied by (if it be not the psychical expression of) certain definite movements of the organs of vision. When we perceive that the relation of *a* to *c* is similar to the relation of *c* to *b*, and is dissimilar to the relation of *c* to *d*, the physiological basis is to be sought in the similarity of eye-movements in the one case, and the difference both in their amount and in their direction in the other.

Once more it should be noted, that although from the psychologist's or physiologist's standpoint space relations may be explicable in terms of transitions in consciousness or of muscular movements, yet from the point of view of the student of physics they are regarded as frankly objective or external to the percipient. The psychologist, from the nature of his analysis, is bound to regard impressions as states of consciousness, and perceptions of relation as transitions in consciousness. But the physicist in his analysis regards, and is wise in so regarding, both impressions and perceptions of relation as representative (it is not for him *quâ* physicist to consider how) of external realities. It is a grave error in scientific method to import into a physical discussion of objective motion in space and time a collateral discussion of psychological questions, still more to attempt a statement of the laws of motion for physical purposes in terms of states of consciousness or the transitions between them.

We may now pass from relations in space to relations in time. Both are alike given, from morning to night, in the daily routine of experience. But both alike remain marginal until reflection turns the focal light of consciousness upon them. Did the reader notice, as he read what was said above with regard to space-relations, that time-relations were lurking throughout in the margin of his consciousness and only needed special attention in retrospection to render them focal? Yet this is undoubtedly so. The impression of the point *a* was *succeeded* by that of *c*. The time relation was subconsciously sensed, but was not made focal in consciousness. As in the case of space relations, so too with time-relations, although we feel them as they pass, yet we cannot definitely perceive them until they have passed. We have in focussing them to look back upon the course of the psychical wave and to fix our attention not upon the impressions as such, not upon the space-relations or the amount and direction of eye-movements in passing from one to another, but upon the sequence or succession. And here, too, it is clear that until the succession *a c* is completed we cannot have an accurate perception of the time-relation involved, or in other words it is only by retrospection that such relation can be focussed.

If we represent diagrammatically, in the accompanying

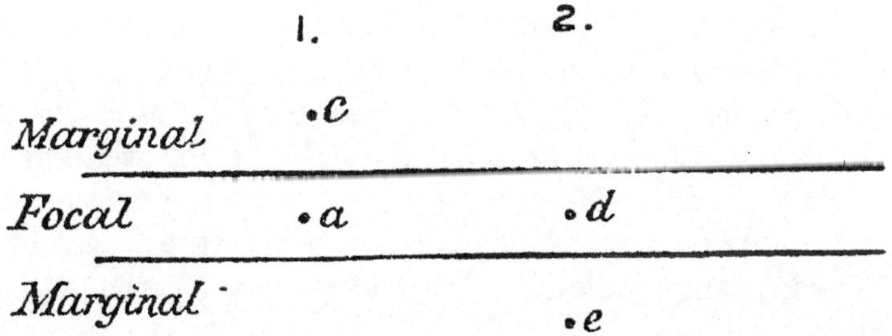

FIG. 16.

figure, the flow of consciousness from left to right with a narrow focal region, and on either side of it a marginal region; then in two succeeding moments of consciousness we have first *a* focal with *c* marginal, and then *d* focal with *e* marginal. When we are directing our attention to space-relations we make a vertical section of the wave and disregard the horizontal flow; and when we are directing our attention to the time-relations we make both vertical and horizontal sections, and perceive *a c* as simultaneous and *a d* as successive.

It is probable that the psychological and physiological basis of the perception of duration and of sequence is the continuance in consciousness and gradual fading of impressions. If impressions and the sensation elements which go to the formation of states of consciousness were strictly instantaneous, there could be no continuity of consciousness and no perception of duration or of sequence. But impressions do not appear and disappear in consciousness instantaneously. They take an appreciable time to fade out of consciousness. If two notes in succession be played or struck on a musical instrument, the first remains marginal in consciousness when the second has come in to occupy the focus. The sound as a psychical experience does not cease abruptly, but gently dies away; and such fading of impressions probably affords the subconscious material of our awareness of duration and succession, and of our subsequent perception of relations in time. And here it is worth noting that our accurate measurement of, and symbolization of time-relations, is through the intervention of perceived relations in space,—the space passed over by the hands of our clocks, the space covered by the movements of the heavenly bodies, or the shadow on a sundial, and so forth.

The raw materials of numerical relations, as of those of space and time, are given in our daily experience, and are marginally sensed long before they are focally perceived. The child, long before he can count, senses the difference between one thing and two things, between two and three, between three and several, between several and many. It would not be surprising to find that a clever dog was able to distinguish from each other playing-cards, from the ace to the ten. But they would be distinguished through difference of sense-impression, not through perception of numerical relations. So, too, with succession. One can very readily distinguish a succession of three from a succession of four, without anything like counting, through the sensing of sense-experience. It is, indeed, surprising how large a group of sounds, up to sixty-four in my own case, can be appreciated as a group without counting. But the perception of numerical relations is something more than the sensing of a group of discrete objects or sounds. It is also to be distinguished from the perception of the group as larger or smaller. Whether the numerical relations were first perceived among objects simultaneously presented, or in association with succession, we cannot say; but it is at least possible, if not probable, that they arose in close association with that phase of time-experience which presents us with succession rather than with duration. Run the eye slowly from left to right along the shaded diagram (Fig. 17). You

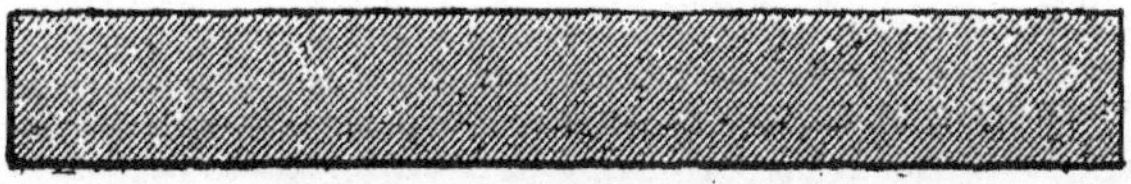

FIG. 17.

are subconsciously aware of the duration of the impression it produces. But if you run the eye along the second figure (Fig. 18), you are aware of succession. The homogeneous

FIG. 18.

duration of a continuous impression gives place to a successive series of similar impressions. And in this series you have, not only one aspect of time-sequence, but also the material from which a numerical sequence may, on the advent of reflection, be evolved.

We have already seen that when the attention is transferred from a red wafer to another red wafer, or from a red one to a blue one, we are subconsciously aware of the similarity or the dissimilarity of the transition in consciousness. For purposes of daily intercourse, and for purposes of physical science, we should speak of the similarity of the red wafers to each other, or of the dissimilarity of the red to the blue. That is to say, we should attach the similarity to the object. But for purposes of psychological analysis we regard the matter from a different standpoint. The red wafer of the plain man or the physicist is for the psychologist a substantive impression, and the similarity of the two red wafers, or the dissimilarity of a red and a blue one, is a transition in consciousness. So long as this transition is only subconsciously sensed, we have merely a dim awareness of the similarity or dissimilarity; but when the transition is rendered focal, we have a clear perception of the relation. And, as in other cases, this perception involves reflection. To perceive the relation between the white and the shaded disc in Fig. 19, we fix the attention first on one and then on

FIG. 19.

the other, so as to make first the white focal with the dark marginal, and then the dark focal with the white marginal. We may then make both discs marginal, and fix the attention on their similarity of size or their dissimilarity in shading, thus rendering, not the discs themselves, but their relation in particular respects, focal to consciousness.

In perceiving such relations there is no necessity that there should be eye-movement. If by means of a lantern a red disc is thrown on a screen, and then the red be suddenly changed to blue, we can perceive the colour-relation, or the transition in consciousness from one to the other. Or, to take the testimony of other senses, we can perceive the similarity or dissimilarity of two succeeding sounds, or of two tastes or scents, or pressures, and so forth. The physiology of these transitions is not well understood; but the similarity or dissimilarity is presumably between like or unlike cortical conditions. Psychologically, what we call a perception of similarity is therefore a recognition, through retrospection, that two succeeding impressions were in certain respects identical, or that the transition was from like to like; while dissimilarity is the perception of a difference in the subsequent impression, or that the transition was from like to unlike. There must be a transition for the perception of similarity or dissimilarity. If there be no transition, we do not speak of the similarity of two impressions, but the continued duration of a single impression. In any case, the essential point to notice is that the definite perception of similarity involves the conscious and focal appreciation of the relation of likeness as such, and differs from the mere subconsciousness or marginal awareness that an impression is one that has occurred before in previous experience. Throw a street urchin a halfpenny, or a dog a piece of biscuit. They do not stop to render focal the similarity of the objects to previous halfpennies or previous bits of biscuit. The one goes for the focal halfpenny; the other for the focal biscuit. Their actions show that they are aware of the likeness of these impressions to previous impressions; but they show no evidence of making the relation focal. If we say, as we do say in popular parlance, that the dog recognizes the object as a bit of biscuit, we must remember that we are using the word "recognize" loosely. And if we continue to use the word, we must be careful to distinguish between this simple and elementary form of recognition, and that higher form of recognition, properly so-called, which implies a conscious perception of the relationships involved. The one is at most a marginal and subconscious awareness that the impression is not new, or is familiar; the other is a fully conscious and focal product of reflective thought. The one is the outcome of sense-experience; the other is a bit of intellectual knowledge.

It is a nice question how far we are justified in speaking of *impressions of relation.* The impression is directly due to stimuli. But when, transferring our attention from a bluish-green wafer to one that is greenish-blue, we perceive the colour relationship, there are no stimuli which give rise to an impression thereof as such. There are special stimuli from the greenish-blue wafer, slightly different stimuli from the bluish-green wafer; but none that we know of accompanying the transition from the one to the other. Nevertheless, I think that it is permissible to speak of the focussed transition between impressions as an impression of relation. We may speak therefore of presentative relations between impressions, and representative relations between ideas, the relation being in either case focal. In one group of relations indeed, those of space, there are motor elements due to the stimuli which accompany the perception of the relation of any two bodies in space as the eyes range from one to the other. And these space relations are those which, beyond all others, are definite and exact. These are they which we use diagrammatically as an aid to the perception of other and subtler relations in the graphic method. And it may well be that it is to the presence of stimuli accompanying the movements that this pre-eminent accuracy is due.

We must now proceed to note that when once the perception of relations has entered into the fabric of the mental synthesis, its results become woven through and through the tapestry of consciousness, so as to constitute an abiding background. When once relations of space, time, number, size, and so forth, have been perceived, representative perceptions may be called up by association, not necessarily into prominence, but into the marginal background of consciousness. The sportsman's recognition of the partridge he brings down is different from the dog's recognition, because for the sportsman there is a subconscious background of relationships which have been perceived, while for the dog this subconscious background is, as I believe, absent. This suffusion of all our impressions with more or less of a definitely relational tone, is what raises them to the level of percepts. For, as I before said, the percept is an impression to which is added a conscious or subconscious perception of relation to the subject or to other objects; or, as we may otherwise phrase it, the percept is an impression set in a relational background. In previous chapters we have intentionally neglected or endeavoured to eliminate all that differentiates the impression from the percept. Now it is necessary to introduce this factor in the mental synthesis, a factor which is, as I believe, distinctively human. Animals and little children live in a world of impressions and ideas, set in a background of dimly-sensed relations which have never been perceived as such. For fully developed man the world is a world of percepts, set in a background of relations which have been consciously grasped. It will be noted, however, that it is not the focal impression itself but its marginal setting which is thus modified. And since we can, in analytic thought, dissociate the impression from its setting, we may perhaps fairly hope that what we have already endeavoured to establish concerning impressions, in sense-experience or consentience, is nowise invalidated by the fact that in us the setting is different from that which forms the subconscious background in the infant and, as I shall hereafter more definitely suggest, in the animal.

We will take one more example of a relation, the definite perception of which differs markedly from the marginal awareness which is its precursor in mental development; and this relation we may term that of association. There are a thousand and one occurrences in our daily life which afford a basis in experience for the perception of such a relation. Numberless events are associated with definite consequences. The occurrence of any one of these begets a feeling of expectation of that which is its normal associate. The child sees a sweet, and as he stretches out his hand for it there rises in his mind, through association, an idea of its taste. The taste is suggested by the visual impression, and *we* say that the two are associated; but for the child there need not be, and probably is not, any perception of the association. The daily life of many of the lower animals is moulded by certain practically constant associations in experience; thus it is they obtain food, thus they avoid danger. But there is no reason to suppose that they can or do perceive these associations as such. Relationships, even the most constant—those by which experience is continually guided—are beyond them, because it is only through reflection that the relationship can be perceived. The most that we can say is, that through sense-experience they are dimly and subconsciously aware of these constant associations. They sense where they cannot perceive.

Enough has now been said in elucidation of the distinction between the focal perception of relations and the mere marginal awareness of the transitions in consciousness from impression to impression. It is of the essence of experience as given in our daily life that substantive impressions stand out like clear-cut figures in the midst of a number of more or less dim and shadowy spectres in the marginal region of the conscious field. And it is just because there is in consciousness not only a focus but also a margin, any spectre in which may subsequently become a focal and clear-cut figure, that we sense or are aware of the relations which the figures and spectres bear to each other. So long as our interest is centred in the figures themselves, their relations to each other are merely incidental. But if we wish to describe the conscious scene, *then* we are forced to bring the relationships into focal prominence. No description, still less any explanation or knowledge, properly so-called, is possible, except in terms of the relations which the figures and spectres bear to each other. To describe, to explain, to understand a scene, one has to dwell upon it reflectively and note the relations which its several parts bear to each other. Try to describe the simplest visual scene, or a commonplace sequence of events, and see if you can do so without using words which render focal the relationships involved. An infant or a dog cannot understand the simplest possible description, because the words which stand for the relationships have no meaning. The words which stand for impressions may have suggestive value. The word "cat" or "rats" may have for the dog a very definite suggestive value; and hence some people fancy that when they say to their dog "there is a cat in the garden," the animal understands what they say. But there is no evidence that the dog understands. It is quite sufficient to suppose that the word "cat" has suggestive force, all the rest being for the dog mere surplusage of sound. Those who have had experience in teaching tricks to dogs or other clever animals will be well aware how futile it is to describe or explain what they are to do. We will not, however, pursue further at present the bearing on animal psychology of what we have learnt in this chapter. But the fact that we cannot describe, still less explain, without rendering the relationships explicit and focal, justifies the suggestion that the perception of relationship, rendered possible by reflection or retrospection, was, in the evolution of man, rendered necessary for the purposes of social communication.

The mode of origin of the perception of relations, is admittedly one of the most difficult steps in psychological development to explain and render probable from the evolution standpoint. The explanation here offered does not lay claim to be in all respects adequate and complete. But I believe it is on the right lines. In the simplest experiences of daily life there are given what we may term, by anticipation, space-relations and time-relations, and arising out of these, relations of size or extent of space-occupancy and of duration or extent of time-occupancy; numerical relations and relations of normal association; relations of colour, tone, and illumination among visual objects; of pitch, intensity, and timbre among sounds; of sweet and bitter among tastes; of pungent, aromatic, and so forth, among scents; of varying pressure and temperature, and all the other differences, among sense-data; and in the midst of all these, relations of similarity and dissimilarity. All of these, and, as life becomes more complex, other less simple relationships are given, and may eventually, in the course of mental development, be perceived. And my object has been to show that, as primarily given in unreflective sense-experience, they remain marginal in consciousness, but that eventually they may become focal through the perception of relations.

In conclusion it must be noted that the relations which we perceive are in all cases particular relations, though it does not by any means follow that the related terms should be particular. The particularity lies in the relation, not in the terms that are related, one or other of which may be a product of that conceptional thought which we shall consider in the fifteenth chapter.

CHAPTER XIV

DO ANIMALS PERCEIVE RELATIONS.

WHEN I take a walk across the downs with Tony, a fox-terrier pup, I carry with me a stick; for it is his delight to race after it and bound back with it in his mouth. The other day I took with me a heavy-knobbed stick, a Kaffir knob-kerrie. At first he seized this by the middle; but to carry it thus was an awkward, lop-sided, unbalanced operation, and by the close of the afternoon he had profited by an hour or two of experience, and seized the stick near the knob end. Now such a proceeding can be completely explained in terms of sense-experience. The process was throughout one of trial and error; gradually he found the most comfortable way of carrying that stick, and adopted it. Incidentally he was solving in a practical way a problem in mechanics; he was finding the centre of gravity of the stick. Incidentally, too, he gave me an opportunity of perceiving that the centre of gravity had certain space-relations. It lay within about seven inches of the knob-end of the stick. But is there any reason to suppose that Tony perceived this relationship in even a rudimentary and indefinite way? I could see none. Through sense-experience he became aware in a practical way of how best to deal with the stick. There is no necessity for the adequate explanation of all that I observed to suppose that the pup perceived the relations as such; the relations at most may be regarded as implicit in practical performance by certain activities, not as explicit in focal perception. If therefore the canon we have already laid down is to be adopted, namely, that in no case is an animal activity to be interpreted as the outcome of a higher psychical faculty, if it can be fairly interpreted as the outcome of one which stands lower in the psychological scale,—if, I say, this canon is to be adopted, then we are bound to interpret the action of the dog as performed through sense-experience alone. And it will conduce to clearness if I here distinctly state that in *sense-experience* and in *intelligent* adaptation to circumstances, there is no perception of relations. This absence of perception is part of the connotation of these terms as I employ them in this work. That is to say, if the dog does perceive relations, then he displays something more than sense-experience and intelligence.

But is not the distinction, on which so much stress is here laid, after all a somewhat trivial one? What does it all come to but the distinction between, on the one hand, the marginal awareness of certain transitions in consciousness, and, on the other hand, the focal and clear-cut perceptions of the *same* transitions? Obviously, I myself think the distinction an important one, or I should not lay such stress upon it. But, of course, if my critic sees no distinction worth the drawing between an undefined marginal awareness and a well-defined focal idea; no distinction worth the drawing between passing on in consciousness from impression to impression and reflectively directing the attention to the nature of the transitions between them; no distinction worth the drawing between the *naïve* reception of a sequence of experiences and the beginnings of introspection, there is an end of the matter between us. We must agree to differ. It certainly appears to me that the distinction between hazy and marginal elements in consciousness and clear-cut ideas is a real one; that the distinction between merely sensing a transition and focussing the attention upon it is a valid one; and that the distinction between *naïve* experience and introspection which involves reflection, is not a matter of mere psychological hair-splitting. And of this I am convinced, that if once the transition from sense-experience to reflective introspection be admitted as one not beyond the possibilities of natural development, the comparative psychologist may rest assured that the natural genesis, through process of evolution, of mind in all its phases up to the very highest, is an hypothesis the validity of which can no longer be seriously questioned. If once the perception of relations and the beginnings of retrospection be granted as possible by natural process of mental development, the key of the evolutionist position is won.

To return to the experiment with the fox-terrier pup and the knob-stick, I repeat that, so far as I am able to judge, all that I observed was capable of interpretation on the hypothesis that the dog was guided in its actions by intelligence, profiting by the associations established in sense-experience. And now I would ask, Of what practical service would it be to the fox-terrier pup to make the relations focal in perception? I am unable to see that it would be of any practical service or advantage to him as a fox-terrier; and being of no practical service or advantage to him, I am unable to see what grounds we have for supposing that this faculty has been developed in him or in his race. And where, it may be asked, is the practical service and advantage of the perception of relations to primitive man, whose rude needs and free wild life may have been in early times but little different from those of the brutes around him? I reply that the perception of relations is a necessary factor in the evolution of descriptive intercommunication. It is a sheer impossibility to describe, not merely to indicate by gesture or by practical demonstration, but to *describe*, where to seize the knob-kerrie, so that it may be comfortably balanced, without perceiving or rendering focal the space-relations. I would ask the reader to try to do so; for his inevitable failure will aid him in understanding what I mean.

When I said just now that the perception of relations, as such, would be of no service or advantage to the fox-terrier pup, I assumed that among dogs, and animals generally, the power of descriptive intercommunication has not been evolved. That they have some powers of indicative communication, both by gesture and by the emission of sounds, is not likely to be called in question by any one who has observed animals; but I am not satisfied that there is any sufficient evidence that animals have powers of descriptive intercommunication. Let us briefly consider the powers that they unquestionably have, and contrast them with the powers the possession of which by animals is doubtful.

I have already briefly described the sounds emitted by young chicks, and stated my opinion of their suggestive value. They afford a good example of one type of indicative intercommunication. They are indicative of emotional states. This type of intercommunication is of wide prevalence and of great importance in the animal world. The song of birds is an elaborated example of such intercommunication. It is indicative of certain emotional states, and I think unquestionably suggestive of answering emotions. And among birds in which class we have this elaborated example of intercommunication through sounds, we have also an elaborated example, in love-antics and display of plumage, of similar indicative communication by gesture. There can be little doubt that the dog, both by sound and gesture,—the tail being a most important organ of communication,—is able to indicate the nature of his emotional states. And such powers of indicative communition are doubtless to many animals of marked service and advantage in the state of nature. I will quote an example, which appears to me a pertinent one, from a pamphlet by Mr H. B. Medlicott:—"In the early dawn of a grey morning," he says, "I was geologising along the base of the Muhair Hills in South Behar, when all of a sudden there was a stampede of many pigs from the fringe of the jungle, with porcine shrieks of *sauve-qui-peut* significance. After a short run in the open they took to the jungle again, and in a few minutes there was another uproar, but different in sound and in action; there was a rush, presumably of the fighting members, to the spot where the row began, and after some seconds a large leopard sprang from the midst of the scuffle. In a few bounds he was in the open, and stood looking back, licking his chaps. The pigs did not break cover, but continued on their way. They were returning to their lair after a night's feeding in the plain, several families having combined for mutual protection; while the beasts of prey were evidently waiting for the occasion. I was alone, and though armed, I did not care to beat up the ground to see if in either case a kill had been effected. The numerous herd covered a considerable space, and the scrub was thick. The prompt concerted action must in each case have been started by the special cry. I imagine that the first assailant was a tiger, and the case was at once known to be hopeless, the cry prompting instant flight, while in the second case the cry was for defence. It can scarcely be doubted that in the first case each adult pig had a vision of a tiger, and in the second of a leopard or some minor foe."*

This interpretation leads us on to the second type of indicative intercommunication,—namely, where in addition to, and perhaps eventually apart from, the suggestion of an emotional state, there is an indication of a particular impression or idea which gives rise to the emotional state. There can be little doubt, I think, that the first stage of indicative communication is purely emotional, and that on this is grafted the indication of particular objects. If, for example, in the vast majority of cases the cry which prompts instant

* "The Evolution of Mind in Man," footnote, pp. 25, 26.

flight among the pigs is called forth by a tiger, it is reason able to suppose that this cry would become indicative, as Mr Medlicott suggests, of the vision of that particular animal. But if the second cry, for defence, was prompted sometimes by a leopard and sometimes by some other minor foe, then this cry might be suggestive of a vision of any one of these, of a succession of visions, or might remain of purely emotional import. Whether the dog barks in different tones to indicate "cat" or "rat," as the case may be, it is difficult to say. But it is possible that he may do so. In any case there can be no question that the dog and other animals respond differentially to different indicative words as addressed to them by us, and therefore presumably have different particular ideas suggested through association with the different sounds.

The investigation of the question, how far, in their intercommunication with each other, separate sounds are suggestive of particular ideas, is one in which experiment, with the aid of the phonograph, in menageries, or in the experimental stations which may some day be established for the accurate study of zoological psychology, may be expected to throw light. Mr R. L. Garner has done pioneer work in this subject by his study of the sounds emitted by monkeys. His conclusions are by no means definite, and his work * contains much hasty and immature generalisation. Of the nine sounds made by capuchins, not one is, so far as the observations go, indubitably indicative of a particular object of desire. All of them may be, and would seem to be, in the emotional stage, and expressive of satisfaction, discontent, alarm, apprehension, and so forth. Still they may be indicative of particular objects of appetence or aversion; and experiments with the phonograph, conducted with due care and under test-conditions, may do much to throw light upon an interesting and important problem.

* "The Speech of Monkeys," by R. L. Garner.

Concluding so far, then, that animals have powers of indicative communication, which is primarily suggestive of emotional states, and secondarily (and probably only incipiently) suggestive of particular objects, we may now pass on to consider the kind of evidence there is for the possession by animals of powers of descriptive intercommunication. There are a good many cases on record of dogs or cats that have been cured, or treated kindly, subsequently returning with a sick or sorry companion. Such a case occurred at the Bristol Infirmary a few years ago. Many a time have I seen Toby, who was always on the alert, invite the fatter and lazier Ginger to accompany him on a desirable worry. In this connection I may perhaps be allowed to reproduce a case I have given in a previous work.* It was communicated to me by my friend, Mr Robert Hall Warren, of Bristol. "My grandfather," he says, "a merchant of this city, had two dogs, one a small one and another larger, who, being fierce, rejoiced in the appropriate name of Boxer. On one of his business journeys into Cornwall he took the smaller dog with him, and for some reason left it at an inn in Devonshire, promising to call for it on his return from Cornwall. When he did so, the landlord apologised for the absence of the dog, and said that, some time after my grandfather left, the little dog fought with the landlord's dog, and came off much the worse for the fight. He then disappeared, and some time afterwards returned with another and larger dog, who set upon his enemy, and, I think, killed him. Then the two dogs walked off, and were no more seen. From the description given, my grandfather had no doubt that the larger dog was Boxer; and, on returning home, found that the little dog had come back, and that both dogs had gone away, and, after a time, had returned home, where he found them."

Such cases as these may be interpreted as involving

* "Animal Life and Intelligence," p. 344.

descriptive intercommunication. It is assumed that the one dog described in some rudimentary fashion to the other what had taken place, or what he had seen, or what he had suffered. We are not at present considering the intelligence, or more than intelligence displayed; we are only looking at the communication as such. And I think it is clear that in none of these cases need there have been more than what I have termed indicative communication. There need not have been any descriptive communication at all. At the instance of certain suggestive signs or actions the one dog follows the other. That is all that is necessary to account for the facts. In accordance with our canon of interpretation, therefore, we are bound to assume no more without further warrant. If, when a little dog had been bullied by a cur, it ran to its big friend; if, after the latter had been induced to follow him, the little dog were removed, and *then* the big dog went and thrashed the cur (following the lines of a well-known experiment of Sir John Lubbock's, with ants); that would afford strong presumptive evidence of descriptive communication. In this, as in so many other cases, what we need is experimental investigation. At present, however, there is not, so far as I am aware, any definite evidence that animals possess powers of descriptive intercommunication involving the perception of relations.

Of the stages by which indicative communication passed into descriptive communication we are ignorant. But for this passage it was necessary that in addition to sounds or other signs symbolic of impressions, there should be sounds or other signs symbolic of relations. And this step which would thus lead to descriptive communication would also lead to knowledge, which implies and is dependent on the perception of relations.

In a suggestive pamphlet on "The Evolution of Mind in Man," Mr H. B. Medlicott, accepting the teaching of those who contend that knowledge consists of relations, sees clearly that the utterance of sounds with relational import was essential as a preliminary step to the development of knowledge. "Relations," he says,* "are immaterial, ultra-physical; the senses cannot touch them directly, and so no sensation of them can be recorded in the brain; and hence, as we have seen to be admitted, no knowledge, in the proper sense of which relations are the secret, could appear. This is the function performed by language, to supply a material notation whereby the brain can take hold of the unseen world of relations, and so convert experience, as unconsciously recorded in "mere consciousness," into knowledge; introducing mind into the organism, changing the brute into man. This is the upward step we were in search of, and it is wholly within the range of natural evolution."

If, as I gather to be the fact, Mr Medlicott regards these relations as something wholly new to consciousness, we must look upon his view as incomplete. It assumes that it is possible for "language to supply a material notation," for what, so far as consciousness is concerned, has as yet no existence. The relations must be already *there*, implicit in sense-experience, before they can become explicit through the instrumentality of a sound or sign by which they may be indicated. But if, as I have endeavoured to show, the raw materials of relations be scattered up and down throughout the whole range of sense-experience, then the step that was effected when relation was first perceived was the rendering focal and clear to consciousness of what was hitherto only marginal and subconscious. And in this contention it is clear that I am not opposing the view to which Mr Medlicott has given expression; I only seek to complete it.

Let me illustrate. A dog has before it two similar objects, one nearer, the other farther. Now the state of conscious-

* Page 24.—In this quotation I have substituted the word "knowledge" for "intelligence," since Mr Medlicott uses the latter word in a different sense from that adopted in this work.

ness he has when he looks at the farther object differs from the state of consciousness he has when he looks at the nearer object; for he reacts differently according as he springs at one or the other; and unless he is an unconscious automaton, the difference in reaction is the index of a difference in the states of consciousness preceding the reaction. But he does not make this difference, which he merely senses in sense-experience, focal as a perception of the relation in space which *we* term distance. Or if he does, he thereby goes a stage beyond sense-experience. The difference is, however, something really present in sense-experience, and is the subconscious determinant of the exact nature of the reaction. In the absence of such elements in sense-experience, the observed phenomena of the perfecting of skill in animals are inexplicable. And the step we are now considering, the step from sense-experience to the perception of relations, is the making focal in perception of this difference as such, which has hitherto been only a subconscious determinant of action in the practical life of sense-experience. The relation itself is not something wholly new to conscious experience; it is only newly perceived as a relation.

To the question—Where would be the advantage of perceiving relations as such, since practical sense-experience suffices for all the needs of existence of many highly organised animals? I again reply that such perception, in association with signs or sounds indicative of the relations so perceived, would render possible descriptive intercommunication, the immense advantage of which among animals living together in a community cannot for one moment be doubted. And if it be admitted that the raw materials of the relation to be perceived are present subconsciously in sense-experience, I join with Mr Medlicott in contending that the eventual bringing them to the focus, and extending the process of indication by signs from impressions to these new focal elements in consciousness, "is wholly within the range of natural evolution."

From what has been said it will be seen that I am disposed to regard descriptive intercommunication as closely bound up with the initial stages of the perception of relations; and I should regard the possession of powers of such descriptive intercommunication in any group of organisms as conclusive evidence that they were able to perceive relations as such. I do not think that such evidence is at present forthcoming with respect to any animals. But it may well be asked whether this is the only evidence that would be convincing. May not relations be perceived by some of the higher animals, and not yet utilised in this way, through the affixing to them of sounds significant of the relations so perceived? I think it not impossible that they may; and though the evidence of the fact may be difficult to obtain, that is certainly no valid reason for discontinuing the investigation. I am, however, firmly convinced that in this investigation the results of careful experiments are of more value than chance observations, perhaps unwittingly touched up into anecdotes. And I would here earnestly request those who are interested in zoological psychology, not to rest content with merely recording an observation of animal intelligence, but, if possible, to make it the basis of such experiments as may help to reveal its true psychological nature. Again and again I receive communications of, or see recorded in books or newspapers, instances of animal activity, which afford admirable points of departure for experimental investigation, and for endeavouring to induce the animal to exemplify the intelligent or other activity under slightly varying circumstances. It is also essential, so far as possible, to watch all the stages of the evolution of any clever trick; for in many cases it is quite impossible to get at the true psychological import of a complex and nicely adapted activity, unless we know something of its embryonic stages.

One of the relations among objects which animals are sometimes supposed to perceive is the numerical relation. Animals are supposed to count. Lichtenberg is reported by Thompson * to have found that a tame nightingale could reckon up to three. He daily gave the bird three meal worms, after which it expected no more, and this not from satiety, for if he enticed it by a fourth, it at once jumped down to receive it. I feel quite convinced that sense-experience was sufficient to enable the bird which had constantly received three worms, to be aware when his daily allowance had already been given him. As has already been said, it is quite possible to sense fairly large numbers, both of objects simultaneously presented (illuminated by the electric spark) and of sounds successively presented. A group of five objects forms a different impression from that produced by a group of four objects or of six objects; and these different impressions can be distinguished in sense-experience. We must not be misled by the fact that we *call* them four, five, or six, and thus use words which imply that *we* have perceived numerical relationship. Sense-experience is too, I think, amply sufficient for the explanation of the sagacity of the suspicious rook described by Leroy. If the rook saw a man with a gun go into a hut, she kept away from her nest. "To deceive this suspicious bird, the plan was hit upon of sending two men into the watch-house, one of whom passed on while the other remained; but the crow counted and kept her distance. The next day three went, and again she perceived that only two returned. In fine, it was found necessary to send five or six men to the watch-house in order to put her out of her calculation."† Here it is quite sufficient to suppose that the rook sensed the impression of two men as different from the impression

* "Passions of Animals," p. 29.

† "Animal Intelligence," Eng. Trans. Quoted by Mr Romanes, "Origin of Human Faculty," p. 57.

of three men, but was unable to sense the difference between the impression of a group of four men from that of a group of five men.

At the instigation, and under the supervision of Mr Romanes, the keeper in charge of the chimpanzee, Sally, at the Zoological Gardens, made this matter the subject of experimental investigation. The facts elicited are of great interest, however we interpret them. This highly intelligent animal was taught to hand to the keeper any number of straws up to five, and subsequently further* on being asked. The straws were first placed in her mouth, and then handed to the keeper. During my visits to the Gardens, I have seen her perform this operation sixteen times, of which eleven gave correct results. But on one day when she was twice wrong (giving three for five, and four for three), the keeper said she was tired, and inclined to be sulky. I did not see any experiments beyond five. Here we have (1) an association of an appropriate group with a given sound, and (2) a distinguishing of the group we call three from the groups we call two, four, and five. Whether the groups were sensed as similar and dissimilar, as the case may be, or were numerically perceived, is a matter of interpretation. My own view is that the sensing of the similarity or dissimilarity of the groups or sequences is not beyond the power of sense-experience, and that there was no true perception of numerical relations—relations which even among the lower members of the human race are only perceived in a very rudimentary fashion, if indeed in the case of some savages they are perceived at all. But this is merely my own opinion. It is possible that, under the guidance of the keeper, Sally was acquiring the rudiments of reflection and perception.

I will now proceed to describe some experiments I have made with dogs the object of which was to ascertain how

---

* See Romanes; *Nature*, vol. 40, p. 160.

far they perceived the nature of a difficulty they had to overcome, or how far they were led to overcome it by the process of repeated trial and error, which constitutes the method of intelligence. I may introduce these experiments by quoting a description I have elsewhere * given. Some years ago, I was out with a gentleman who was teaching a couple of Scotch terrier pups to carry sticks. Each had a light cane in his mouth. After a while we came to a gate, at the side of which was a gap for foot-passengers between two uprights. We passed through and watched the puppies. Both blundered against the uprights, which caught the ends of the canes. There was a little scrimmaging, and some further ineffectual struggles, and then both dropped the sticks and came through. Their master sent them back to "fetch." The first to arrive at the gap just put his head through, seized a cane by the end and dragged it after him. The other ran through the gap, picked up the cane as usual by the middle, and blundered as before. Again he dropped it and came through. I then went back and placed the stick, so that he could put his head through and seize the end as the other had done. But again he went through bodily, picked up the cane by the middle as before, and blundered. Then his master tried to teach him how to do it. On our return an hour or so afterwards I held the first pup, so that it might be seen how far the other had learnt his lesson. He blundered, however, as before. Then we called him off, and allowed the other pup to have his turn. He, too, blundered for a little, and then came back to us. We passed through the gap and called him after us. Again he blundered; but then, dropping the cane, came through, and, turning, seized the cane *by the middle*, and tried to pull it after him. Of course it caught, and fell out of his mouth. He then seized it nearer the end.

* *Fortnightly Review*, August 1893, p. 223.

Even so it caught; but, by turning his head about, after some little scrambling, he eventually pulled it through.

These pups, then, did not act alike; both had to learn by experience to meet new circumstances. It is remarkable that the apparently more intelligent pup when sent back in the first instance seized the cane at once by the end and dragged it through; and if the observations had been carried no further, one might have supposed that he clearly perceived the best means to effect the desired result. But the second time he did *not* seize the end of the stick, and this may well lead one to suppose that it was rather good fortune than clear perception which made him successful before.

Since writing the above, I have made this matter one of experimental investigation. I will describe the results I obtained with Tony, the fox-terrier, then about fourteen months old. The scene of operations was a field along one side of which run vertical rails about six inches apart, between which he can readily pass. There is one place where a rail is absent, and the gap is therefore twice the usual width. Along one side of the field, at right angles to that described, there is an ordinary open iron fencing.

First day.—Standing on the path adjoining the field, and separated from it by the first mentioned vertical rails, I sent the dog after a short stick into the field, and called him back through the railings. The stick caught at the ends. I whistled, and the dog pushed and struggled vigorously. He retired into the field, lay down, and began gnawing the stick. I called him, and he came up slowly to the railings, and stuck again. After some efforts he put his head on one side, and got the stick, a short one, through. I patted him, and showed him my satisfaction. Then I sent him after it again. He came up to the railings with more confidence, but, having the stick well by the middle, found his passage barred. After some struggles he dropped the stick and came through. I sent him back to fetch. He put his head through, and seized the stick by the middle and then pulled with all his might, dancing up and down in his endeavours. Wriggling his head in the efforts, he at last got the stick through. A third time he again stuck; again dropped the stick; and again seizing it by the middle tried to pull it through. He failed, and came to me. I put the stick so that he could seize it by one end and pull it through. But when I sent him after it he went through himself, picked it up by the middle, and tried to push his way through, succeeding, after many abortive attempts, by holding his head on one side.

Second day.—I sent him after a quite short stick about nine inches long. He came back holding it at one end cigar-fashion. It struck a rail, and he put his head on one side, and came through easily. Sent after it again, he seized the stick by the middle and stuck; but soon got through by holding his head on one side. After repeating this several times, he came up to the railings with his head sideways. I sent him a dozen times after this stick, that he might gain experience of how to deal with the difficulty. I then sent him after a longer stick. He brought it by the middle, holding his head on one side; it caught, and he struggled. Then he dropped it and came through. I sent him back. He again seized it by the middle, and, approaching the railings as before, tried to push through; then, failing, dropped the stick, and could not be induced to make a further attempt. I took him for a walk of a mile or so, and on our return sent him through where the rail was missing, and, standing just there, called him. He came up with some little force, and the rails caught the ends of the stick. He dropped it and came through; and on being sent back seized it by the middle, and tried to get it past by hard pulling. I sent it again into the field. He raced about with it, and came to the open railings; went through and ran down the path to me. I threw it again just through the railings, and called him towards the wider gap; the stick caught, and he then raced round through the open fence. Twice more he did the same thing. After that he ran round at once without trying the vertical railings.

Third day.—Experiments with sticks at first gave the same results as before. The dog had not in the least improved in bringing a long stick through the railings. But he had learnt by experience that he could easily run round with the stick, and would not try to bring it through. He had the intelligence to shirk the difficulty which he could not overcome.

After some weeks I tried him again. The stick caught by the ends and he pushed with great vigour and assiduity, but with no result. He then dropped it, and jumping over it put his nose through the railings, and catching it by the middle tugged for three minutes, dancing up and down in his energetic efforts. I placed one end near the wider gap and induced him to take it in his mouth, and pull the stick through. I then threw the stick again into the field, but he caught sight of the open railings (I had purposely approached the field from the other side) and ran round.

On a subsequent occasion I prepared a short yew stick with a crook at one end where a side branch had been partly cut off. I brought the dog up to the field on the side furthest from the open railings and sent him through after this yew stick. It caught; and after many attempts to push it through he dropped it. I sent him back, and after a while induced him to take it by one end and drag it after him so that it might catch by the crook at the other end. He then tugged at it in the most ridiculously energetic fashion. Nothing could apparently be simpler than to push the stick up, free the crook, and pull the whole through; but the dog continued to pull. I repeated the experiment many times, and tried to show the dog how the difficulty could be overcome. But each time the crook caught, he pulled with all his strength, seizing the stick now at the end, now in the middle, now near the crook. At length he seized the crook itself, and with a wrench broke it off. A man who was passing, and who had paused for a couple of minutes to watch the proceedings, said, "Clever dog that, sir; he knows where the hitch do lie." The remark was the characteristic outcome of two minutes' chance observation. During the half-hour or more that I watched the dog he had tried nearly every possible way of holding and tugging at the stick. And such is the method of sense-experience—continued trial and error until a happy effect is reached. I subsequently tried Tony with a similar crooked stick, but he had spied the open railings, or remembered them, and ran round with it and came down the path.

I have experimented with several other dogs in the same way and with similar results, to describe which would be mere wearisome repetition. I will therefore content myself with adding a few words concerning an hour's observation of a very intelligent Yorkshire terrier, Toddy. This dog is exceedingly bright, active, and clever at such tricks as finding a ball or other object hidden in his mistress's drawing-room. His master, my friend and colleague, Professor Leipner, came with me and stimulated Toddy to his best efforts. I do not propose to give the results in detail; they were quite similar to those with Tony. The general results may be thus summarised. It was only by chance that the dog seized the end of the stick and pulled it through. At the end of the hour he was no better than he was at the beginning. When the stick was placed in the best position for seizing one end and pulling it through, the dog seized it by the middle, walked with it into the field, and brought it up to the railings. The effect produced on both our minds was that Toddy showed no evidence of perceiving the way in which the difficulty could be overcome.

Now I am particularly anxious that it should be quite clearly understood that I do not for one moment doubt that, with a little trouble and patience, one could teach a dog—or that he could teach himself—to bring a stick through railings. Dogs can be taught, and may acquire through their own efforts, much more difficult and complex tricks. What the observations show, so far as the dogs observed are concerned, is that their way of dealing with the difficulty is the method of trial and error, which is the method of sense-experience. In other words the facts observed can be completely explained on the hypothesis that there is sense-experience only. The perception of relations as such is not necessary to the performances, and is therefore by our canon of interpretation excluded.

I have experimented with dogs in the following way. I have tied a thread to a stick or other object with which they would play, and have drawn the object towards me so that there should be established an association between the movement and me—the movement being always in the same direction. I have then passed the thread round the leg of a chair, or some such support, and caused the object to move off in the opposite direction. I wished to see if there was any evidence of surprise at a result which was quite new to the dog's experience. I have not succeeded in observing anything of the sort. Mr Romanes, however, found that one of his dogs showed surprise and disquietude, rising at last to dread, when a dry bone with which it had been playing on the lawn was made to come to life, and given supernatural powers of activity by means of a thread.* I have tried a similar experiment on a good many dogs, but with no positive result. One young and timid pup did indeed show fear : but he showed similar signs of fear when any thing was thrown or rolled towards him.

* "Fetishism in Animals," *Nature*, vol. xvii. 168.

I have several times noticed that when a bone is tied to a piece of fine string, and a dog made to race round after it as it is swung in a circle, the dog will after a while bite at the string and so avert the motion of the bone. And it may be said that such an action implies a perception of the relation which the string bears to the bone in motion. I am sure, however, that such perception is not necessary. The dog sees the bone swinging round, and sees the string. After many ineffectual attempts to seize the bone which he cannot catch, he at last snaps at the string which he can catch. That he senses the string as partaking of the motion of the bone cannot be questioned; but I do not think there is any necessary perception of the relations as such. Nor indeed is the moment of excited activity one in which a relation as such is likely to become focal. In our own moments of strenuous action, though we may and do profit by the results which have been wrought into our nature by the perception of relations, we have no time to make them focal. Such focussing of relations is the result of reflection. It can hardly be said that a dog who is introduced to something so new to his experience as a bone flying round and round at the end of a string, has had any previous data for the quiet perception of the relations involved.

The conclusion to which I think we must be led by the foregoing considerations is that, in answer to the question which heads this chapter: Do animals perceive relations? we must reply that all the ordinary activities of animals can be explained on the supposition that they do not. Furthermore it would appear that experimental observations tend to support the view that sense-experience is all-sufficient for them, and that in face of an unwonted obstacle they trust entirely to the method of sense-experience, that of trial and error, for dealing with it. But it will doubtless be said that it is not in such cases that we are forced to admit a higher faculty than that which falls under the designation sense-experience. It is in the remarkable cases of "the reasoning power of animals," that we find a body of evidence which forces upon us the conviction that they are rational beings. Before considering such cases, however, we have to learn how the perception of relations leads up to conceptual thought and to reason; and we have to make it quite clear in what sense we are to employ the words "reason" and "rational." It is not a matter on which those who have attentively studied psychology are by any means agreed. And whether animals have the faculty of reason or not very largely depends on the exact sense in which this word is used. The psychologist is apt sometimes to smile when after the recital, probably in the correspondence columns of a newspaper, of some anecdote of animal intelligence the writer exclaims, If this is not reason I do not know what reason is. As, however, in such cases the writer has himself suggested the alternative, there is perhaps no discourtesy on the part of the psychologist in accepting it.

CHAPTER XV.

CONCEPTUAL THOUGHT.

IN foregoing chapters I have endeavoured to show how the step may be taken from sense-experience, in which the transitions from impression to impression are merely sensed as they occur in the margin of the conscious field, to the definite perception of relations, where the transitional relationships as such become focal to consciousness. We there saw that perception takes its origin in close touch with practical experience. The relation as perceived is between impressions so recent that they have not yet faded from the margin of consciousness. Take, for example, the spatial relation between two adjacent points. They are still in the field of view, and were but just now focal as impressions. That is to say, when the relation between them is perceived, they are still marginally present to consciousness as the points between which the relationship holds good. And the relation is, as we have seen, between the impressions of concrete and particular objects, or between such a concrete impression and an idea.

Now when once relations between concrete impressions or images have been clearly perceived in the course of practical experience, the materials are given for a further mental process. Two particular objects, billiard balls for example, are perceived to be a span asunder. The balls are removed, and pieces of chalk being substituted they too are seen to be a span asunder. All sorts of small objects may be substituted for the chalk, and no matter what the particular objects may be, the space relationship remains unchanged.

The relation persists throughout all changes of the things which exhibit the relationship. And as we begin to realize this, we are able to think of the relationship of distance, irrespective of the objects which exhibit the relationship—to think of it as something that will apply to a great number of particular cases which, in other respects, may show all sorts of differences. The relationship is no longer particular to those two billiard balls or those two pieces of chalk, but applies generally to all sorts of objects which have been, or may be, met with in experience. We neglect all that is variable and focus the attention on the uniform relation in space. We do not merely *perceive* those two billiard balls to be related in such a way, but we *conceive* the relationship in its abstract and general form. We have reached a conception, and this conception is not concrete, particular and individual, but abstract, general, and of universal application.

The process of conception involves an extension of that reflection which we have seen to be essential to the process of perception, out of which it arises. It is only when we can look back upon the results of a good deal of experience that the permanence of the relation, amid many modes of its manifestation, begets the general conception. In the course of experience we see that this man is like that man, this tree like that tree, this apple like that apple, this coin like that coin, and so on in numberless particular cases. So long as the likeness is particular, we perceive the similarity. But when, in reflection on numerous particular instances, we see that all possess this relationship in common, we reach the conception of similarity. We are no longer considering *particular* transitions in consciousness, but are reviewing a number of transitions which have occurred in the course of our experience, and which are seen to have a common character. Suppose that we are looking over a series of coins or other similar objects. As we pass from one impression to another, we sense, or are marginally aware of, the similarity of each to each. We may then make the likeness of any two focal in consciousness and perceive the relation of similarity. We may go further and perceive that the relation of this to that is similar to the relation of that to the other—we may perceive the similarity of the relations. But the relations that we perceive to be similar are particular relations. Not until the particular fades from view, and the relationship, as common to all the particular instances, becomes focal, do we reach the conception properly so-called.

It will be seen that this process by which conceptions are reached involves abstraction and involves analysis. The power of abstraction has its germ in sense-experience. It is essentially a concentration of attention on the focal element to the neglect of marginal elements in consciousness. When in the primitive life of sense-experience the attention is so rivetted on some focal impression that the margin almost fades out of view altogether, we have the process of abstraction in an incipient phase. For abstraction, to a large extent, is a matter of the relative intensity of the focus and the margin. In perceiving the relation between particular impressions and ideas—between the colour shades of two skeins of wool for example—it is essential that they should be held representatively in the margin of consciousness as the terms between which the relation is perceived to hold good. But when we focus the relationship of similarity in general, and form a conception thereof, we no longer hold any particular terms representatively in the margin of consciousness. We think of the similarity as focal to a marginal field into which any pair of an indefinite number of similar terms may be introduced. Whether we could do so to much purpose, in the absence of a name or some such symbol for the focal idea thus generalized, is a question which is open to discussion. I am inclined to think that we could not. Our conception of similarity is indefinite from its very generality, and the moment we try and make it clear and precise to our mind's eye, we make it particular by thinking of an illustrative example. We exemplify the conception by reference to a particular perception. The symbolic name, however, serves to fix the general conception without particularizing it. It enables us more completely to abstract the focus from the indefinitely variable margin. And all this involves analysis. We are not merely looking back upon our past experience and reviewing it in memory; we are analysing it, and making a new synthesis out of the results of our analysis. We find in a great number of particular cases, with which reflection presents us, the relation of similarity, and submitting these cases to analysis, we detach the relation from the related terms. But the relation is given in experience as a similarity now of colour, now of musical notes, now of pressures, now of tastes, now of scents, and so forth. Fusing these together, we reach the synthetic general conception of this relation as of universal application, and label it "similarity." And we should not fail to notice that this process of analysis and synthesis, as leading to the conception of relations, is a fully conscious process. No doubt the mere repetition of particular cases of the perception of similarity would beget a sense of familiarity, a fringe of marginal awareness of repetition. But this, so long as it continues marginal, could not rise into conception. For conception involves a definitely conscious grasp of a synthetic unity as the result of reflective analysis. At the risk of wearisome reiteration, let me again say that in primitive sense-experience we may have a practical awareness of similarity, such awareness being altogether marginal; that when this marginal awareness comes to the focus in early stages of reflection, the similarity between particular impressions may be definitely perceived; that mere repetition of, and marginal familiarity with, such perceptions will not generate conceptions, which are the definitely focal results of abstraction, analysis, and synthetic generalization.

Let us now pass on to consider the conceptions that we frame of what we term the qualities of objects. I say advisedly "what we term" the qualities of objects, because at the stage of primitive sense-experience the qualities of objects are not yet distinguished. And though, in describing and endeavouring to explain sense-experience, we are forced to use separate words for the qualities that *we* distinguish, it must be remembered that at this stage of mental development the distinction has not yet been made. For the animal—or let us say, to avoid controversy, for such animals as are still in the stage of sense-experience—there is no distinction of quality and object; there are merely focal impressions and ideas, set in a subconscious margin. Suppose that a puppy is gaining experience of things which are good to eat. Associations are established between certain visual impressions and certain gustatory impressions. *We* speak of the puppy as dealing with objects which have the quality or property of edibility. But the "objects" and the "edibility" are our affair, not the puppy's. Or let a child be playing with red and blue marbles, he may sense the similarity of the red marble to the blue marble in the matter of form, and their dissimilarity in the matter of colour. But when once these relations begin to be perceived, the form and the colour begin to be distinguished. Qualities begin to emerge, and to be dissociated from the continuum of sense-experience. But the qualities as perceived are merely the salient features of certain given impressions; the dissociation is only incipient. This red rose, as an impression, is distinguished from that yellow rose, and is assimilated to the other red rose. The quality of colour predominates over that of form, but the two are still associated together, and associated with the scent which the sight of the rose suggests. For at this stage we are still in close touch with practical experience; and in practical experience there is no dissociation between colour and form, nor between the scent of a rose and the rose which is scented. But when reflection deals not merely with the data of experience as they are practically given—in their very presence, so to speak—but ranging over a wider field, analyses them, abstracts their focal essence, and generalizes the results in conceptual synthesis, then, and not till then, does the quality dissociate itself in thought from the object. The rose, the violet, and the carnation are given in sense-experience as differently scented; perception comes upon the scene, and the scent relations of the flowers are perceived, still in close association with the other qualities which are presented to consciousness in practical experience; finally, conception analyses the results of experience, isolates a particular quality already predominant in perception, and lays it before us as the named quality of "scent." And so pleased are we with the result of our analysis, and with the distinction it has enabled us to draw, that we no longer say that the rose or the carnation *is* scented, but that the flower *has* scent; attributing to the object the "possession" of the quality which we have thus reached as a general conception.

When once such conceptions have been reached, they tend to suffuse and modify our whole mental outlook. The case here is analogous to those to which attention has already been drawn. We learnt that, so soon as the perception of relations has entered into the fabric of the mental synthesis, its results become woven through and through the tapestry of consciousness, so far as to constitute an abiding background. And we saw that this suffusion of all our impressions with more or less of a relational tone is what raises them to the level of percepts—the percept being an impression set in a relational background, the relations in which have been at some time or other definitely perceived as such. Similarly, so soon as the general conception has entered into the fabric of our mental synthesis, its results too become woven through and through the tapestry of consciousness, so as to constitute an abiding conceptual background. And just as the percept is an impression set in a background of particular relations, so is the concept an impression which is known to be but a particular manifestation of that which is of general or universal import.

Mark well the social value of the concept. Perception, although it probably took its origin in close connection with the need of intercommunication, is nevertheless in the main an individual matter. I perceive the particular relationships of this, that, and the other object with which practical experience brings me into contact. I may describe to you what I perceive, and hear your description of what you have perceived. But it is not till we each of us, you and I, rise above the particular perceptions to general conceptions, that we begin consciously to participate in our common humanity. The percept at best leaves us units with powers of intercommunication; but in the concept we merge our individuality, to share a common nature.

Note, too, the increasing complexity and the increasing richness of the fabric of the mental tapestry as we proceed from sense-experience to perception, and so to conception. In sense-experience we have impressions and their revived ideas, set in a dim background of a mere awareness of the transitions which take place at this stage of mental development. Association has these impressions and ideas to deal with, and nought beside. But when relations have been perceived, a new set of ideas is introduced, and association weaves these strands into the more complex tissue of consciousness. And when relations and the qualities of objects have not only been perceived, but generalized so as to acquire widely extended meaning, these general conceptions are utilized by association, and the tapestry of consciousness reaches all the wealth and richness which the interlacing of the strands afforded by sense-experience, perception, and conception can give it. And this wealth and complexity concerns not only, I think we may say not chiefly, the focus of consciousness; it is wrought into the marginal background. When one reaches the climax of a fine play or a great novel, it is not merely the focal impression or idea of the moment that stirs us, but the extraordinary richness and intensity of a state of consciousness embracing representatively a multiplicity of duly subordinated details.

For beings who have reached the conceptual stage then, association deals with a new order of ideas,—those of general import. And it is now time to revise and restate our definition of the term "idea." We began by defining the impression as that which is rendered focal to consciousness through afferent impulses, and the idea as that which is rendered focal to consciousness through the revival or the representation of such impressions. But when we reach the perception of relations, we have a new order of ideas,—the ideas of relation; they are the revivals of those perceptions which, since they were generated in close touch with the related impressions, we termed impressions of relation. And when we rise to conception yet a third order of ideas is reached,—those of general or universal application. It may be well to distinguish and differentiate these three orders of ideas by writing the word without italics or capitals when the ideas of sense-experience are referred to; by using italics for *ideas* of or involving particular relations; and by using a capital letter for Ideas in their general sense. Thus I may have an idea of a snake as an object of sense-experience; an *idea* of its length; and an Idea of its zoological position. The word object is another term that is used with a very wide range of signification. We speak, in the first place, of an object of sense-experience when we say, for instance, that a puppy has an impression of the object that we call a bone. Secondly, we speak of perceiving an object as related to another object in space. Here the object hitherto merely sensed is perceived in certain of its relations. This is the object as percept. Thirdly, we speak of an object with reference to its general purpose, of a clock, for example, as a timepiece. This is the object as concept. And again, we speak of an object in the logical sense as the synthesis of the qualities which we have analytically distinguished. This is the logical object. Thus when, for purposes of logical thought, we define an "object," we enumerate its several qualities in varying relationship to other objects and to ourselves. And though it may be said that the term as employed of sense-experience and of the earlier stages of perception is used, by anticipation, for that which, in the light of conception, is known as an object,—since an object, it may be said, is only known as such in antithetical relation to the subject,—still, for reasons that will hereafter appear, it is convenient to retain all these uses. With regard to the logical object,—that is, the synthesis of the qualities which we analytically distinguish,—we may note that there are certain properties or qualities which are essential to the object as a synthetic unity; these are termed primary or essential qualities. And there are other qualities which are variable or unessential; these are termed secondary or accidental qualities. Take, for example, such an object as a teaspoon. There are certain properties or qualities, weight, resistance, and a definite form, in the absence of which the object would not be a teaspoon at all. But it may be hot or cold, silver or plated, fiddle-patterned, scalloped, or plain, and still be a teaspoon. These are accidental qualities. Of course, as actually presented, this teaspoon now in my hand—this object as percept—is hot, made of silver, and scalloped. Nothing can alter that as a fact of experience. But these accidental qualities are not essential to that general Idea of a teaspoon, which constitutes the logical object.

This digression on the use of terms leads us on to consider the connection of the faculty of conception with words and language. It has been pointed out in previous chapters that the step from sense-experience to perception, rendered possible by incipient reflection, probably had its origin in the needs of intercommunication. It is impossible to describe a simple visual scene or a common daily occurrence without using terms which stand for the particular relations involved. But the terms, though they primarily express particular relations, stand for any such relations. They are thus generalised, and so far conceptual in their nature. Hence, if it was the need of intercommunication which gave rise to perception, it was also the means of communication which largely facilitated the further process of conception. And we may well believe that perception and conception had their origin in tolerably close association with each other. It is not my purpose to speculate here concerning the origin of language, and how it played its part, vital as that part undoubtedly was, in mental development; it will be sufficient to indicate the relation of the name or word to the impression, the percept, and the concept. In the first place, the word may be associated by contiguity with an impression, and may therefore serve to revive the corresponding idea. The infant tolerably early gives evidence of the establishment of such associations; and a clever dog will show by his actions that a considerable number of words call up in his mind the ideas of the impressions with which they have been associated. And these words are not merely associated with revivals of impressions of the special senses, they also serve to suggest activities. The dog lies down, sits up, begs, fetches and "drops it," as he is told. What we may term the first stage of naming, arising out of the indicative stage of intercommunication, is therefore the association of a more or less arbitrary word-sign or symbol with some sort of impression. The second stage is the association of such a symbol with a relation definitely perceived. By this means we express particular characters of the impression, in relation to other impressions, particular qualities of objects in relation to other objects, or to the percipient, particular modes or intensities of the activities. Such associations are considerably later in development in the human infant; but it is exceedingly difficult to determine just when they are formed. The child grows up in what we may term an atmosphere of language. At first this environment of language is all mere sound; but associations of particular words with particular impressions are soon established, and we then say that the child is beginning to understand what is said to it. It is certainly some little time, but one finds it hard to say how long, before the relational words begin to have their true significance. If only we could remember these early stages in mental development, what a boon it would be to psychology! As it is, we are left to the conjectural interpretation of observed activities. And we find that it is easy to get a response in answer to the suggestion, "Baby, laugh," but difficult to get a different response for "Baby, laugh loud," and "Baby, laugh softly." The words for perceptions or their *ideas*, since they are expressive of relations, are relative to other words, and they serve to define the relationship of impressions and their symbols to other impressions and their symbols. The impression-words are for the most part nouns and verbs; the relation-words are adjectives, adverbs, and prepositions. These are all relative to each other. There is not an adjective nor an adverb nor a preposition that does not more or less clearly imply such relations. Many of them go in couples, such as "hard" "soft," "near" "far," "swiftly" "slowly," "to" from," "up" "down," "above" "under." But when the relation is not so obvious, when, for example, we say that the grass is "green," even here the relationship is none the less implied; for colour is not *perceived* at all (though it may be sensed) until its relationship to other colours is grasped. In saying that the grass is green, as a matter of definite perception, we imply that it is not red nor brown nor blue, nor any other colour than green.

The third stage of naming is the conceptual stage, and this involves the generalization of that which has been acquired in and through perception. And this is not so much the formation of new names for general and abstract relationships,—though many general and abstract names are then first given such as "redness," "colour," "distance," "opacity," "thing," "virtue," and the rest,—it is not so much, I say, the naming of these general and abstract Ideas, as the raising of all the rest of language from the particular to the general, and the consequent viewing of all that is named in the light of conception. Common nouns for objects of experience such as "dog," "chair," "tree," "rock," or oft-repeated activities such as "to run," "to call," "to touch," are then no longer of only particular application. They stand no longer for simple impressions, no longer even for particular percepts, but they are henceforth symbolic of concepts. The word "ball" which for the child is suggestive merely of a particular impression or idea, is for the man who lives in a conceptual atmosphere symbolic of this and of much besides, it not merely stands for this particular ball as actually sensed and perceived, but is symbolic of a general concept which is but exemplified in the particular object before us. The name of the general concept thus embraces thousands of particular objects which are, so to say, samples of that which is so named, or exemplifications of the Idea so symbolized.

I said just now that the words which are associated with impressions are nouns and verbs; but the verb in such connection must be emptied of its predicative value. For predication is beyond the reach of sense-experience. The word "is," or its representative in the verb as predicative, implies the exercise of perception or of conception. It is indeed primarily symbolic of the transition in consciousness from an impression to some perception or conception which the impression suggests. And it enables us to state in the form of a proposition the transition which has thus taken place. Thus it comes to be symbolic of all those transitions in consciousness between focal states which are perceived or conceived to be related. As I look up from my page I see the dog, the rug, the fireplace, and so on. These, as merely successive impressions, are not related, and I can make no affirmation concerning them. But in the light of perception I affirm that the dog *is on* the rug and *in front of* the fire. The impressions are brought into connection with each other through the perception of certain spatial relations, and the transitions having a new value given to them, are symbolized in predication. When, in Brazil, I saw coiled on the road before me a brilliant black and red snake, the sequence in consciousness was an impression, clearly defined in vision, followed by the concept I symbolize as coral-snake. And I indicated this transition by saying to my neighbour, "That *is* a coral snake." Then followed further transitions in my consciousness which I indicated by adding, "It is terribly poisonous, and has recently cast its skin." The first of these transitions was from the general concept coral-snake to one of the qualities involved in the concept. The second was suggested by a peculiarity in the appearance of the snake. And when my companion asked me why I supposed that it had recently cast its skin, I replied, "Because of the exceeding brilliance of the colours of the snake."

That little word "because" introduces us to a group of words which are essentially characteristic of the conceptual phase of mental development. They are symbolic of logical inference and explanation. We may *describe* events and experiences by the hour, keeping close to the plane of perception (though the words we employ are full of conceptual import), and never have occasion to use one of these words, such as "hence," "because," "therefore," and the like; but the moment we seek to *explain*, the moment we have to answer the question "Why?" they come into prominence and symbolize a particular way of regarding some of the transitions in consciousness. Let us take the case of the inference concerning the shedding of the skin by the coral-snake as an example of such transition, and let us suppose that my companion was one of those worrying people who will get to the bottom of everything. He is not content with the explanation I give him of the transition when I say that I feel sure that the snake has recently cast its skin because the colours are so bright. He asks on what grounds I am so confident in the matter. I reply that I have good reason for confidence because having seen a good many snakes of various kinds it has been my invariable experience that colours so bright, and of that peculiar freshness, are only presented by individuals which have recently exuviated. But still my friend is not satisfied. He inquires whether I have seen many coral snakes which have just shed their skin. I confess that I have never had opportunities of observing snakes of this particular kind. By what right then, he urges, do you extend to snakes of this kind conclusions you have drawn from observations on other kinds? I remind him that I am not relying on my own observations only, but upon those of many well-qualified observers. Granted, says my friend, but by what right do you extend to this particular snake the results of observation on any number of other individuals? Because snakes are all alike in this respect, I reply. But how on earth do you know that? Neither you nor any one else can have examined all snakes that exist and have existed; and if you had, how can you be sure about the snakes that at the time of the investigation were still unborn? Well, I reply, perhaps a little nettled, notwithstanding your pretended doubts, I shall continue to assume that all snakes are alike in this respect until I find that my assumption is incorrect. Very good, but forgive my insistency, why on the grounds of a paltry thousand or two of observations do you propose to assume that all the millions of snakes that are in existence now, and shall hereafter be born, are alike in this respect? Because, I reply, nature is uniform, and the universal testimony of science proclaims the fact. But suppose you were to find a snake to-morrow that had just cast its skin, the surface of which was dull and blotchy, you would no doubt abandon your assumption of the uniformity of nature. Not a bit of it, I reply, I should feel certain that the poor thing was in ill health or diseased, and that there was good cause for the anomaly. One more question, says my friend. You hold, I believe, that all snakes have ribs. Now suppose that in Central Africa there were found a new group of snakes, not a single diseased individual, mind you, but a family of snakes with several genera and species, which were quite destitute of ribs, then at least you would abandon your assumption of the uniformity of nature? Not so, I answer; if they were without ribs, they would not be snakes as at present defined; and if they were proved in all other respects to be true snakes, we should have to alter our definition of snakes. We used to believe that all mammals brought forth their young in a fairly advanced state of development, but now we know that the duckbill of Australia lays eggs. We have had to alter our Idea of the mammal, but that does not show that nature is not uniform. I see, says my friend with a smile, that this assumption is so deeply rooted in your mind that nothing is likely to shake it. Of course it is, I reply; *it is the condition of all explanation;* for if nature be not uniform what is the good of my trying to explain anything, since the explanation will be invalidated by departures from uniformity? And what is the good of forming a general concept of anything, such as this which I define as a snake, if its qualities are variable through the variability of nature? All explanation depends on uniformity in that which is to be explained, and uniformity in the use of the terms by which it is explained.

I have thus led up to a conception of wide range and generality—that of uniformity—in the light of which certain transitions in consciousness are explained, and have indicated its three-fold applicability. If our terminology be not uniform—if "coral snake" and "poisonous" sometimes mean one thing and sometimes another—we cannot express the logical sequence of our thought. If our concepts be not uniform—if "coral snake" sometimes includes the factor poisonous and sometimes does not—our thought will have no logical sequence to express. If nature be not uniform—if coral snakes are sometimes poisonous and sometimes harmless—our thought and its expression may be logical, but this will not enable us to explain the world in which we live. Uniformity of symbolization (one word one meaning),* uniformity of thought (the meaning nowise variable), and uniformity of nature (experience as trustworthy), these form the triple support of the "therefore;" these are the conditions under which alone we can attempt explanations of ourselves or of our surroundings. Of these three, the uniformity of nature is altogether beyond our control. If experience is not trustworthy, that is not our fault. But if we fail to keep the concepts clearly defined and invariable, or if

* To prevent misconception, let me add that in saying "one word one meaning," I intend it to be understood that, for purposes of exact logical thought, every term must *pro hoc* be rigidly defined, and must not be used in any other sense than the one so defined. In the general usage of language, the greater number of the words we employ are in themselves indefinite (*i.e.* centres of divergent association), and become definite only in and through their environment in the context. The life and freedom of language would be lost if all words were bound down by rigid definition. Language would then cease to be organic, and become merely mechanical, as it is, indeed, in purely technical scientific statements, and in the syllogism of formal logic.

we use the same word in several different senses, that is the fault not of nature but of our thought, or of our mode of expressing it. And such faults are technically called fallacies. If a man tells us that the part can be greater than the whole, we know that he is talking nonsense. Either his conceptions or his words, or both, fail to conform to the conditions of uniformity. Either his conception of "whole," or his conception of "part," or his conception of "greater," or the terms by which he expresses one or more of them, are different from mine. If a man tells us that though all coral snakes are poisonous yet this coral snake is harmless, he contradicts himself. And if in explanation he tells us, with a knowing wink, that in this coral snake the poison fangs and glands have been extirpated, he does not one whit the less contradict himself. His statement involves a fallacy since he uses the word "coral snake" in two senses. What he ought to say is, that though all unmutilated coral snakes are poisonous, yet this mutilated coral snake is harmless. That is a perfectly logical proposition. If a man, to take but one more example, tells us that an object can be at one and the same time above and beneath the table, he is talking nonsense. The conception above is one thing, the conception beneath is its antithesis. In affirming that it is above, he is implicitly denying that it is beneath; and in affirming that it is beneath, he is denying that it is above. He cannot affirm that it is both above and beneath without a contradiction; and a contradiction is a statement in contravention of the uniformity of thought. Statements, the opposite of which involve a contradiction or breach of uniformity of thought, are called necessary truths. That a thing cannot both be and not be at the same instant of time, that things which are equal to the same thing are equal to one another—these are necessary truths. The conclusion of any syllogism the premisses of which are granted, is a necessary truth. All such necessities are logical necessities or necessities of thought.

In formal logic, the premisses and the conclusion are stated in a syllogism, of which the following is a type:—

This is a coral snake;
All coral snakes are poisonous;
Therefore this is poisonous.

In such a syllogism (which is here stated in its most convenient psychological form, rather than its formally deductive order) we rise at once from the particular impression, or perhaps percept, to the logical concept ("This is a coral snake"). We then set forth one of the factors of the concept ("All coral snakes are poisonous," or, better, "The concept coral snake involves the conception poisonous"), and then proceed to apply the general proposition to the particular case, saying "therefore this is poisonous." This conclusion is necessary, because it is already implicit in the concept, and only needs to be rendered explicit through the syllogism.

It must be clearly understood that such explanation, whether expressed in the syllogism or otherwise, is something more than, and something different from, a description of the mental process as it actually presents itself at the moment of experience. When my eye fell upon the snake lying coiled on the Brazilian road, what took place in my mind may be thus expressed. Black and red object—coral snake—poisonous. Or, in general, impression—concept—one salient factor in the concept. And again, while the concept "snake" was still present in the background of consciousness, still forming part of the state of consciousness as a whole, there was the transition:—Bright colour—recently cast skin; a transition from a particular perception to a suggested antecedent of similar perceptions. In each case we have a suggestion through association. I then said to my companion, "That is a coral snake; it is poisonous; and has recently cast its skin." This was a condensed and incomplete statement of what had passed through my mind.

When I was asked the grounds on which I made the statement, I replied in effect, "That is a coral snake, and *therefore* poisonous. I know it has recently cast its skin, *because* of its brilliant colour." I thus laid stress upon and rendered focal the nature of the transitions as of that particular kind which we term logical inferences, and symbolized their nature by the words "therefore" and "because." The further questions of my companion were directed towards the justification of these inferences, and this led us to the general conception of uniformity. But at the moment of experience this elaborate justification was not present to consciousness. And I am not a little anxious that it should be quite clear that the experience itself, the passing of the mind from coral snake to poisonous, or from bright colour to shed skin,—what we ordinarily call jumping to the conclusions,—is one thing, and its logical justification quite another thing. There is many a boy who, as the result of his own observation and what he had read or been told, might have jumped to the conclusion that the coral snake had recently cast its skin, but who, when pressed for the grounds of his conclusion, would not be able to get further than, "Snakes always look like that when they have just shed their skin;" and when asked on what grounds he supposed the snake to be poisonous, would say, "Coral snakes always are,"—the demand for further explanation serving only to confuse him.

It is very essential that this important difference between the occurrence of certain transitions in consciousness, and their logical justification, should be clearly grasped; for loose thinkers are continually confusing the two. A child sees a sweet and eats it. We explain this as a case of association. The sight of the sweet, we say, suggests its edibility. But did not the sight of the coral snake at once suggest its venomous nature? And if we explain the latter case in logical form—this is a coral snake, and therefore venomous—why should we not do the same in the case of the infant, and throw that too into logical form—this is a sweet and therefore good to eat? And since such logical expression involves a well-defined concept for a mediating term, how can it be denied that the child who eats the sweet has formed a concept? This is scarcely an exaggeration of the mode of procedure of certain loose thinkers. The fallacy, however, becomes patent when once the fact is grasped, that the sequence through association (coral snake—poisonous) is one thing, and its logical justification (coral snake *therefore* poisonous), involving as it does a clear perception of the relation symbolized in the word "therefore," is another thing. There is no evidence that the infant can think the "therefore." When we further remember that we symbolize by the same word "sweet," both the sense-impression which the child experiences, and the concept which is the product of reflective thought, we shall see that the expression: Sweet—good to eat, may be used of the child to describe the sequence of an impression, and an idea in sense-experience without implying either perception or conception.

I am particularly anxious that this distinction should be quite clear. Sense-experience and association afford the basis of a great number of expectations of the greatest practical value in the conduct of life. Such expectations may be described as intelligent inferences or inferences in the field of sense-experience. A dog suddenly scents a trail, or spoor, which suggests through association the idea of a fox. We may say that he intelligently infers that a fox has passed that way. Sambo lying on the rug in the hall, sees his master come downstairs in a black coat, watches him put on a tall hat, and pick up his gloves, and is content to let him depart; but if he comes down in a slack coat, the dog is on the alert; and when his master puts on his felt hat, Sambo's joy knows no bounds. The intelligent inferences may be thus symbolized,—

Black coat and tall hat—no walk.
Slack coat and felt hat—walk.

A boy who has seized a stinging nettle, and made his fingers smart, will carefully avoid touching a dead nettle. He intelligently infers that this leaf which is so similar to that in appearance, is like also in its stinging properties. The inference is false, and may subsequently be corrected by further experience. It is an expectation generated through association by superficial similarity. A very large proportion of our practical inferences in daily life are of this intelligent type. Such inferences are sometimes said to be due to the "unconscious logic of common sense." The phrase is, I think, unsatisfactory and misleading. There is no logical process, conscious or unconscious. These are intelligent inferences due to association which may (or may not) be explained and justified by logic. Nor is the phrase "implicit logic" any more satisfactory. The logic is not implicit but absent. What is present is the raw material on which logic may or may not be exercised. The relations are implicit; not the logic by means of which they may be focussed.

To this process of practical inference through association in sense-experience, the term *reasoning* is applied by many writers. It is sometimes spoken of as reasoning from particulars. Professor Sully terms it implicit reasoning. There is, however, a narrower use of the word reasoning. This may be described as the process of passing on from proposition to proposition, the connections (transitions) between which involve, *and are seen to involve*, the logical relation symbolized by the word "therefore" or "because." Reason, in this narrower sense, implies the power of perceiving, and conceiving the logical relation as such. This is the sense in which I use the words "reason" and "reasoning," reserving the word "intelligence" for the faculty in virtue of which inferences are suggested in the field of sense-experience. On this view, if we wish to determine whether an inference is rational or intelligent, we must inquire whether the logical relation is clearly perceived or not. That being alone is rational who is able to focus the *therefore*.

The materials on which reason, in this more restricted sense, exercises its function are thickly strewn along the path of our daily experience, and were present for ages before reason came on the scene of mental life to explain them. Sense-experience affords the raw material out of which all our higher conceptual thought is elaborated. But only in the light of reflection are the relations which are involved perceived and conceived as such. Only in the light of reflection does the logical connection of the sequence of conscious states emerge into clear view, and does the faculty of reason become manifest. The function of reason is to explain, to disclose, and set forth the relations. And this it may do through the mediation of the syllogism. This schematic arrangement of the propositions, however, I repeat, does not describe and is not intended to describe, the manner or order of transition at the moment of experience. It may chance to present that order, or it may not. What it does is to place the propositions in convenient sequence for logical exposition and justification.

And when through reflection and the perception of relationships, we rise above the plane of sense-experience, when to associations by contiguity or by superficial resemblance, there are added associations through perceived similarity of relations, the process by which inferences are actually reached differs markedly from that by which they are logically justified to the eye of reason. The Idea shoots into the mind by some subtle association by relational similarity. The man who has this power in marked degree is said to have the gift of insight. Of such a nature are the inspirations or flashes of insight of the artist, the philosopher, or the man of science. For the man of science, at his best, is an artist, whose art-work is the interpretation of nature. As artist, he sets forth his Idea in due subservience to the laws of artistic proportion. But as reasoner he has to justify his conception. He has to set forth its logical sequence and relations. In doing this he is not describing how he came by the Idea. It may have flashed across his mind, as the hypothesis of natural selection flashed across the mind of Darwin, when he was reading his Malthus. He may himself be quite unable to say how it came. The artist seldom knows ought of the genesis of his Ideas and ideals, and attributes them to inspiration, rating severely the psychologist for attempting any explanation. But it is not the office of logic to say how the Idea originated. Logic justifies the inspiration, and exhibits its relation to the rest of our thought. Logic is thus the afterthought to insight.

We must now proceed to note that all our thought and its logical expression are ultimately based and founded in experience; and thus we come to the uniformity of nature (experience as trustworthy) as a condition of thought. It is only on the assumption of the uniformity of nature that we can rise from the particular percept to the general concept. If experience be not trustworthy, we can never profit thereby. A trust, conscious or unconscious, in the uniformity of nature is not only a condition of all explanation and interpretation of natural operations; it is a condition of all mental development. The dog that snaps up the bone which is thrown to him exhibits a quite unconscious trust in the uniformity of nature. The whole range of intelligent actions in men and animals is conditioned by the uniformity of nature. But it is only by slow and painful steps, involving generalizations that are successively wider and more comprehensive, that a conception embracing the whole of nature as definitely uniform is reached. Reason then looks back upon the ladder by which she has climbed, and then understands how it has been possible for her to climb at all. The uniformity of nature is a condition of all mental development, as of all reasoning, but it is not conceived as a condition until reason is firmly settled on her throne. And when reason is thus seated, even then we may jump to conclusions a hundred times a day without being once called upon to give a logical justification. For it is one thing for the mind to pass from this phase of consciousness to that, and a different thing for reflective thought to give a logical justification of the passage.

The uniformity of nature by which all our mental development has been conditioned, has thus begotten our uniformities of thought, and conversely our uniformity of thought has reacted on our experience of nature, and rendered the conception not merely of wide generality, but universal. By no inductive process can we prove the universal; but reason, assuming the uniformity of thought, universalizes the uniformity of nature. Does it sound somewhat extravagant to say that thought endows nature with this universality of law, this all-embracing linkage of cause and effect? What then are cause and effect in their empirical aspect? Are they not the succeeding phases of certain invariable sequences? Do they not sum up for us certain very general or universal relations? But what are relations? Have we not seen that they are transitions in consciousness? Thus only, at any rate, are they known to empirical psychology. Does it not, then, inevitably follow, that the causal nexus, as relational, is something which our thought impresses on its raw material of sense-experience? But this raw material of sense-experience, we must remember, contains implicitly that which may become explicit through perception and conception. And it is this implicit objective character, and nought else, which, in the light of reflection, becomes explicit as objective uniformity. Thus the objective and the subjective so act and react on each other as to suggest that in their inner essence they are but one. And there we must leave the question, till we have occasion to take it up again in a future chapter.

CHAPTER XVI.

DO ANIMALS REASON?

WE are now in a position to consider the question which I have prefixed as a heading to this chapter. It is obvious that before attempting to answer the question we must be quite clear as to what we mean by reasoning. If we apply the term "reasoning" to the process by which an animal, profiting by experience, adapts his actions to somewhat varying circumstances, there can be no hesitation whatever in giving an affirmative answer to the question. But if by reasoning we mean the process of drawing a logical inference; if we define reason in such a way that it is necessary for the reasoning being as such, to think the *therefore*, then we cannot answer the question so readily. Let it be clearly understood that I use the word in the latter and narrower sense. Using it, then, in this narrower sense, I may paraphrase the question with which this chapter deals, and ask—Do animals focus the *therefore?* Do they think the *why?*

Once more I draw attention to the canon of interpretation adopted at the outset of our inquiries concerning other minds than ours, namely, that in no case is an animal activity to be interpreted as the outcome of a higher psychical faculty, if it can fairly be interpreted as the outcome of faculties which are lower in the psychological scale. Our question then becomes: Are there animal activities the performance of which is inexplicable if the animal in question does not perceive the *why* and think the *therefore?*

Tony, the fox-terrier, already introduced to my readers,

when he wants to go out into the road, puts his head under the latch of the gate, lifts it, and waits for the gate to swing open. Now an observer of the dog's intelligent action might well suppose that he clearly perceived how the end in view was to be gained, and the most appropriate means for effecting his purpose. The following chain of ideas might be supposed to pass through the dog's mind, not, indeed, in a clear-cut logical form, but at anyrate in a rough and practically serviceable way: "Why does that gate remain shut? The latch holds it. I'll lift the latch. Now it is no longer held, therefore it swings open." But is it necessary to assume that there were ideas involving, even in the most rudimentary way, the why and the wherefore? May not the action be quite well explained on the hypothesis that the dog acted under the sole guidance of sense-experience? I think it not unlikely that two chance observers, both witnesses of the action, might long discuss it from different standpoints: the one contending that it was probably based on practical experience, and therefore intelligent as I use this word; the other asking, in reply, what practical experience the dog could have had of gates swinging open of themselves when the latch was raised, insisting on the purposive nature of the action, and asserting that it unquestionably involved a clear perception of the relationship of the *how* of the gates opening, and, as it swung back when the latch was raised, of the *therefore* involved. Eventually the two observers might agree to differ, and each go on his way, more convinced than ever of the correctness of his own view, and of the fact that preconceived ideas may blind a man to the most obvious conclusions.

But if, now, one had an opportunity of seeing how such a clever trick originated, if one had watched all the stages of its genesis, then one would presumably be in a better position to offer an opinion in the matter than either of the chance observers who saw it only in its final and perfected form. With regard to this particular trick of Tony's I am in that position. I watched from the first the development of the habit. The facts are as follows: I may premise that the gate is of iron, and has iron bars running vertically with interspaces of five or six inches between. On either side is a wall or low parapet, on which are similar vertical rails. The latch of the gate is at a level of about a foot above that of the top of the low wall. When it is lifted the gate swings open by its own weight. The gate separates a small garden, of only a few square yards area, from the road. When the dog is put out of the front door he naturally wants to get out into the road, where there is often much to interest him; cats to be worried, other dogs with whom to establish a sniffing acquaintance, and so forth. I said just now that I watched the development of the habit from the beginning. This perhaps is a slight exaggeration. I was sitting at a window above the garden, and heard the dog put out of the door. I therefore watched him. He ran up and down the low wall, and put his head out between the iron bars, now here, now there, now elsewhere, anxiously gazing into the road. This he did for quite three or four minutes. At length it so happened that he put out his head beneath the latch, which, as I have said, is at a convenient height for his doing so, being about a foot above the level of the wall. The latch was thus lifted. He withdrew his head, and began to look out elsewhere, when he found that the gate was swinging open, and out he bolted. After that, whenever I took him out, I shut the gate in his face, and waited till he opened it for himself and joined me. I did not give him any assistance in any way, but just waited and watched, sometimes putting him back and making him open it again. Gradually he went, after fewer pokings of his head out in the wrong place, to the one opening at which the latch was lifted. But it was nearly three weeks from my first noticing his actions from the window before he went at once and with precision to the right place and put his head without any ineffectual fumbling beneath the latch. Even now he always lifts it with the back of his head and not with his muzzle which would be easier for him.

With regard to this particular trick, then, I venture to affirm that, *when we know the whole history of it*, Tony's action is quite similar in kind to that of my little chick, Blackie, which, profiting by a chance experience, pulled down the corner of the newspaper and escaped from my experimental poultry-yard. As it stands, it is quite within the range of sense-experience; nay more, it affords a very pretty example of the application of sense-experience to new circumstances. It is typically intelligent.

One of my own students, Mr Edward J. Shellard, informs me of a case which fell within his own observation, of an action which "appeared at first to be the result of thought," but which on closer investigation was clearly seen to be the outcome of sense-experience. A Scotch staghound living in a yard was often shut out, and in order to enter raised the latch of the door opening into the yard. Here again the chance observer would be likely to fix upon the fact of the latch being purposely raised, but the investigator would seek to know the exact manner in which the action was performed, and how the habit was acquired. In this instance the manner was as follows: The staghound "at first raised his paws to the door and scratched violently, manifesting various signs of impatience. His scratches, which extended from the top of the door downwards and over the whole area, would thus inevitably at some time or other reach the handle of the latch, which was thus struck forcibly downwards, the latch itself rising upwards. The door would then open from the weight of the dog pushing against it. The dog always opened the door in this manner from the time when the incident was first noticed until he left, a period of about three years. The door was opened with no greater ease at the expiration of that period than at the commencement. His paws would strike various parts of the door, and he never appeared to exercise any degree of judgment in the localization of his strokes, the fact of his paws striking the handle of the latch being a necessary result, provided the dog had sufficient patience and strength to continue."

Now what I am particularly anxious to enforce is, not the adoption of the usage of the word "reason" in its narrower sense, though I think that this is desirable; nor the hypothesis that animals do not reason in this sense, though I think that this is probable; but rather that what we need is careful investigation in place of anecdotal reporting. Unfortunately the opportunities for investigation are not numerous. One may keep dogs and other animals for years, and find few opportunities of investigating what has even a semblance of reasoning. And if a friend tells one of the clever performance of his dog, the time for investigation may have passed. No one, by observing Tony opening the gate *now*, could ascertain the stages of the development of this intelligent action. Still opportunities do present themselves; and if those who take a scientific interest in zoological psychology will endeavour to utilize to the full these opportunities, and will record the results of experimental investigation, we shall acquire a better acquaintance with the psychological processes in animals than we could gain by a thousand anecdotes. In zoological psychology we have got beyond the anecdotal stage, we have reached the stage of experimental investigation.

Without such investigation the anecdote, however interesting in itself, is of little service so far as the question of intelligence or reason is concerned. Take, for example, the case of a dog which enters a stream at a point higher than that which he wishes to reach on the opposite side, thus allowing for the force of the current. This may be the result of practical experience and an example of intelligent inference.

Or it may involve the perception of relations and a thinking of the *therefore*, and thus be the result of reason. To determine which, so far as it can be determined, we must know the genesis of the habit. If, for example, a dog which has been accustomed to swim in a lake, and has therefore no practical acquaintance with the effects of a stream, comes to a running river, and, seeing sticks and straws floating down, at once and without previous trial takes to the water further up stream than the point he desires to reach, and does so every time he is brought down to that river, then the probability of his reasoning in the matter would I think be strong. It is difficult to see how, under these circumstances he could adopt the course he does without some such thinking of the *therefore* as we should express as follows: Sticks floating in that water are carried down, therefore if I float in the water I too shall be carried down. It may be said that dogs have been observed, in dealing with tidal rivers, to look at the water and the objects floating in it, so as to ascertain which way the current is flowing, and to take to the water in one direction or the other as the results of observation prompted. How valuable would careful investigation of such cases be! But if the dog had practical experience of the flow in either direction, and had opportunities of associating his own translation with the movements of sticks or other objects, this is perhaps not beyond the reach of intelligence. It would be a case of profiting by associations established through practical sense-experience. The instance we are considering is a case of complex skill; and the varied skill of animals, dependent upon sense-experience, is so wonderful and admirable, that intelligence may reach results which are astonishingly complex. If we regard intelligence based on sense-experience as the determining factor in the psychology of animals, we may well expect this factor to be raised to its highest pitch of excellence. Such I believe to be the case. I believe that the higher animals deal practically with their environments just as well as it is possible to do under the limiting conditions of sense-experience. And I believe that the extraordinary difference between men, even the lowest, and animals, even the highest, is due to the introduction of the new factor involved in the perception of relations and conceptual thought, the one in close association with descriptive intercommunication, the other in association with explanatory intercommunication.

I have before stated my opinion that there is no evidence of powers of descriptive intercommunication among animals. I need hardly add that there is no evidence of explanatory intercommunication. The questions *how* and *what*, and the answers thereto, may be regarded as symbolic of the descriptive stage; the question *why*, and the answer thereto, as symbolic of the explanatory stage. But it may be said that it is possible that the *how* and the *what* may take rudimentary form in the animal mind without necessary symbolization. I do not think that this is likely. If one asks oneself how a thing is done, or what a thing is, one is communing with oneself, and such communing is begotten of a desire to perceive the relationships, and belongs to a stage beyond that of sense-experience. Even the vague informal wondering "what it is" or "how to do it," quite apart from any definiteness of question, presupposes the possibility of perceiving the relations in solution of the wondering. When I toss to one of my experimental chicks something quite new to his experience, such as a red-currant, he looks at it with hesitation in his actions. Had he powers of descriptive intercommunication, he might make some sign for *what?* and receive in reply some sounds significant of the nature of the object. But such question would be bred of descriptive intercommunication, and in the absence thereof would not, I think, arise in any conceivable inarticulate form. I remember once, at the Cape, turning over a large ant-ball, and stooping down to pick up a curculio, when something caught my eye which caused me to hesitate and draw back. (It was merely the head of a lizard.) There were two stages of consciousness. First, a mere sense of something strange and unusual; secondly, the cognitive stage expressed in the question, What is it? The first was, I take it, the sense-experience stage; the second the perceptive stage, which involved a would-be perception of the relations of the object. If when I am geologizing I notice a good rock-exposure, but do not at once see how best to reach it, I am not conscious of the question *how* arising in my mind in any form whatever. First one course, then another, suggests itself, and so on till I start off in a particular line. The *how* only arises when I want to describe to myself or to another the most likely route. I do not think there is any psychological equivalent in the mind of the animal of the *how* or the *what*, or if there is, it is the outcome of the incipient perception of relations, and is at a stage beyond that of sense-experience.

Still less is there, in my opinion, any psychological equivalent of the *why*, by which I mean of course the proximate *why* that can be answered by a *because*, not the ultimate *why* of philosophical inquiry. The *why*, like the *how* and the *what*, is the outcome of, or let us rather say the concomitant of, a phase of intercommunication to which the animal has not attained. But further, the *why* presupposes a phase of intercommunication on a somewhat higher level of development than the descriptive phase. That which is sought in answer to the *why*, whether one asks oneself this question or asks it of another, is an explanation, and no explanation is possible until the relations which have been perceived in particular instances are conceived to hold good generally for all like instances. Take the case of lifting the knob-kerrie in such a way that it balances. Tony can do this practically as the result of trial and error in sense-experience. One can describe how to lift it and how to hold it without going beyond the limits of the knob-kerrie; but one cannot explain why it should be lifted thus to produce the desired result without tacit reference to the general properties of heavy bodies and the principles of the lever. One can describe events or occurrences or facts in terms which are not necessarily significant of more than particular relations; directly one attempts to explain them, one must use terms which are significant of general relationships. We describe the fall of a stone to the ground as a particular occurrence; we explain it proximately as an example of gravitation, a general property of material bodies. Without generalization there is no explanation.

But, as has already been pointed out, when once sounds were used as significant of particular relations, these sounds would be significant also of similar relations, and would become generic. The generalization would not be a conscious process, but one that arose inevitably out of the conditions of signification by sounds. Fully conscious generalization—generalization conceived as such—would come much later. The generalization that I speak of as the inevitable outcome of the use of sounds significant of relations, unquestionably holds good also of sounds indicative of impressions or their corresponding ideas. When the pigs, before described in a quotation from Mr Medlicott, made certain sounds which were, in his opinion, indicative of tiger, it was not a particular tiger that through the indicative sound took form as "a vision" or visual idea,—it was tiger. The visual idea was a composite image, and was generalized in the sense in which I am using the word. Nearly all the ideas (that is to say, representations of impressions) of adult life, whether of men or animals, are of the nature of composite images. They are what Mr Romanes has termed *recepts*, and are, as he has shown, generic in character.

At the very outset therefore of the process, however and

whenever it occurred, of using sounds significant of relations, these sounds would have generic force, and would therefore be at once ready for use, not only for description but for explanation. I surmise therefore,—and one is not in a position to do more than surmise,—that explanation, though it involves the higher faculty of conception, had its origin, not indeed coincidently with, but very shortly after the power of description. The two have by no means advanced *pari passu* in the different races of man. The enormous extension and expansion of the tree of knowledge that has its roots in explanation is characteristic of civilized peoples; but there probably is no race on the surface of the earth now, —there possibly has been no race since man was man,— in the members of which the means of descriptive intercommunication has not prompted the beginnings of explanation.

I return now to the question, whether there is in animals any psychological equivalent of the *why*. It is sometimes said that though the animal may not be able to utter in any form the *why*, and though it may not be able to think it in the clear-cut manner in which the human reasoner thinks it, yet it may trouble him in a dim inarticulate informal way. I suppose I might express this view in the phraseology so often adopted in this work, by saying that though the *why* may not be focal, it may be marginal, and may come tolerably near the focus. Does not, for example, it may be asked, the curiosity evinced by animals show that there arise, at any rate marginally, if not focally, psychological equivalents of the *what*, the *how*, and the *why*?

Mr Romanes records an interesting observation of which one of his dogs was the subject, and which, since it is typical of its kind, we may here briefly consider. The dog was cowed by the sound of apples being shot on to the floor of a loft above the stable; but when Mr Romanes took the dog up into the loft, and let him see what was going on, he ceased to be disquieted by the noise. In his index, Mr

Romanes enters this under the heading, "Appreciation of cause by the dog." Now here, as we have seen to be so often the case, two interpretations are possible. If the dog really perceived the relation of causation as such, he had rational grounds for ceasing to be disquieted; but if it was merely a matter of association of impressions and ideas, his satisfaction was simply intelligence. Let me illustrate what I mean by a somewhat parallel case. Some years ago my cat was asleep on a chair, and my little boy was blowing a toy horn. The cat, without moving, mewed uneasily. I told my boy to go on blowing. The cat grew more uneasy, got up, stretched herself, and turned towards the source of discomfort. She stood looking at my boy for a minute as he blew; then, curling herself up, she went to sleep again, and no amount of blowing disturbed her further. With regard to my cat's procedure, and that of Mr Romanes's dog, I suggest the following explanation:—For animals, I take it, common and familiar sounds soon have their normal associations. An unusual sound without such normal associations leads, if loud and sudden, to undefined fear (for example, with my chicks when I played a sharp chord on the violin); if the sound be merely disturbing, to discomfort or uneasiness; if simply unfamiliar, to curiosity. There is also a normal association of sound with some object, and, in the case of an unusual unfamiliar sound, an almost irresistible tendency to make the source of the sound focal to consciousness and to vision. Hence numberless cases of animals being attracted to objects of which they show signs of fear. Mr Romanes's dog and my cat were both disturbed by unfamiliar sounds; when they were allowed to see the source of the sound, and found nothing further disquieting about it, nor suffered any unpleasant consequences, they were satisfied. And subsequently, the sound suggested that which neither inspired fear nor produced uneasiness. That is the explanation on the hypothesis of intelligence.

And do not the facts, whichever way they are interpreted, justify the inference, that there is a vague or marginal, if not definite and focal, *what* and *why* in the dog's mind? First, I think we may quite fairly exclude the *why*, or reduce it to a *what*. It is surely sufficient to suppose there is an inarticulate "what is it?" which is a practical question, and deals with the present, or just past, experience, without introducing the *why*, the answer to which involves conceptual thought, and a reference to the generalization of perceived relations. When the dog heard a strange and disquieting sound, and when the cat had its first auditory experience of the instrument of torture known as a child's toy horn, we need not, I take it, go beyond the *what* to symbolize the mental attitude. And if we use the *what* as symbolic of the mental attitude, largely emotional in its character, which we call curiosity, to deny the *what* would be to deny curiosity in animals, which no one who has observed them is likely to do. But I question the advisability of using the *what* in this sense.

I revert to my chicks. Suppose I separate them into groups. I throw red-currants, objects quite new to their experience, to one group. Let us grant that they view them in the mental attitude of curiosity. Let us, for the sake of argument, grant that we may symbolize this state as a vague *what?* And let us suppose that the chicks can indicate this emotional state—as they certainly can indicate some emotional states—by a distinctive sound; and let us call it the *what* sound. On first seeing the currants they all make the *what* sound; but since none of them has any experience of the object that prompts the sound, it remains an unsatisfied *what*. They then gain experience of the currants, and find them to have a peculiar taste, and so forth. Let us now suppose that they come to indicate this particular object of sense-experience by a sound which we will call the *currant* sound. This, I may say, is probably beyond the power of chicks; but we have seen that animals (*e.g.* Mr Medlicott's pigs) may use indicative sounds in this way. Unless animals have the power of descriptive intercommunication, this is as far as they can go. The *what* sound may evoke in reply a *currant* sound. I now introduce into this group, who have gained experience of currants, one of the chicks from the other group which has no such experience. It immediately utters the *what* sound, and one of the chicks standing near utters the *currant* sound. What does this convey to the utterer of the *what* sound? Nothing at all. It is a sound indicative of certain experiences which it has never had. To a chick which had learnt to associate the sound with a particular object seen and tasted, it might have value. But to the inexperienced chick it is a sound of no indicative value. At the stage of sense-experience, and merely indicative intercommunication, the *what* (except as symbolic of the attitude of curiosity) has no value, and is not therefore likely to arise.

This little parable will serve to indicate by implication that the *what* presupposes the perceptive stage, with its concomitant stage of descriptive intercommunication. If a child asks concerning any object new to its experience, What is it? he does not want a mere name in reply; and if a mere name is given, he will, if his natural tendencies have not been ruthlessly snubbed down, repeat his question more emphatically, and say: But *what* is it? The only possible answer to this *what* is a description, in terms of the experience that he has, of the new object of which he he has at present no experience. Suppose that a child is standing on the pier by the open sea, and observing a large dark animal throw itself out of the water, exclaims, "What is it? "That is a porpoise," he is told. "But what is a porpoise?" "Well, it is something like a fish in shape, but it must come up to the surface to breathe, and its tail is——," and so on. I repeat, that the *what* implies a descriptive answer; that description is impossible in the absence of the perception of relations; and that I do not know of any good and sufficient evidence that there is any psychological equivalent in the mind of the animal of even the *what*, still less of the *why*.

I return now to the direct consideration of the question which heads this chapter, and which may, as I have said, be paraphrased by the question:—Do animals think the *therefore?* There can be no doubt that there are a great number of recorded cases, which have probably been observed objectively with due care and exactitude, and which have been interpreted by the observers in the terms o rational thought. I take an illustrative case from Mr Romanes's well-known collection.* "Mr Stone writes to me from Norbury Park," he says, "concerning two of his dogs, one large and the other small. Both being in a room at the same time, 'one of them, the larger, had a bone, and when he had left it the smaller dog went to take it; the larger one growled, and the other retired to a corner. Shortly afterwards the larger dog went out; but the other did not appear to notice this, and, at any rate, did not move. A few minutes later the large dog was heard to bark out of doors; the little dog then, without a moment's hesitation, went straight to the bone and took it. It thus appears quite evident that she reasoned—That dog is barking out of doors, therefore he is not in this room, therefore it is safe for me to take the bone. The action was so rapid as to be clearly a consequence of the other dog's barking.'"

Now, here a course of action, which is quite readily interpreted on the hypothesis of sense-experience through association, is explained in rational terms, and the dog is supposed to think the *therefore*. But on similar principles every case of association may be thus interpreted. Suppose

* "Animal Intelligence," p. 460.

that I throw to my two-days-old chicks some caterpillars of the cinnabar moth, and that after a few trials they avoid them; this is explicable as a simple case of association. One of the chicks walks up to a caterpillar, and is apparently about to peck, but checks the action, wipes his bill, and walks away. Nothing is easier than to interpret this intelligent course of action in rational and logical terms:—Black and gold caterpillars are nasty; that is a black and gold caterpillar; therefore it is nasty. This does not seem to me to be at all more far-fetched than the suggestion of Mr Stone's that his dog reasoned:—That dog is barking out of doors, therefore he is not in this room. It is perfectly open to anyone to accept the definition of the word *reason* according to which it includes all cases of profiting by experience through association; and he may then with perfect propriety speak of my two-days-old chick and Mr Stone's dog as rational beings. But I submit that he may not assume that a rational being, as thus defined, can think the *therefore*. For him there must be rational beings, or reasoning beings, in untold numbers which are unable to do anything of the sort.

A well-known writer, Dr Andrew Wilson, describes the case of a dog * which used to hunt a rabbit nearly every morning down a curved shrubbery, and each time ran it into a drain at the end. "The dog then appears to have come to the conclusion"—I quote Dr Wilson's words—"that the chord of a circle is shorter than its arc, for he raised the rabbit again, and, instead of following him through the shrubbery as usual, he took the short cut to the drain, and was ready and in waiting on the rabbit when he arrived, and caught him." It is here assumed that the dog perceived the relation between a chord and its arc. I do not myself believe that he did;—but that is not the question;

* Quoted in "Animal Intelligence," p. 461.

it is merely an expression of individual opinion. The question is:—Can we or can we not explain the dog's action as the outcome of sense-experience, as indicative of intelligence profiting by association? I do not see how this can be denied. The terrier used to start the rabbit nearly every morning, and each time saw it escape into the old drain. There were thus ample opportunities for establishing an association between rabbit and drain. That the sight of the rabbit should suggest the drain into which it daily escaped, and that when the idea was suggested, the dog should run there directly, is a sequence not impossible, one would think, to sense-experience. And if so, the canon of interpretation, so often referred to, makes it imperative for us who adopt it, to accept the interpretation of the action as due to the simpler exercise of intelligence based on sense-experience, rather than that according to which the dog perceived the relation between a chord and its arc.

It is difficult to put the matters we are now discussing to the test of experiment, but wherever an opportunity presents itself we should make the attempt. Some years ago I was with a friend watching a horse which was dragging a heavy load, zigzagging up a steep hill. "How well," said my companion, "that horse appreciates the principle of the inclined plane." I begged him to explain exactly what he meant by this expression. "He sees," said my friend, "that the less the slope the less intense, if more prolonged, the labour to raise the load through a given height." I told him that my own view was that the poor animal was quite unable to perceive the relation between the slope and the labour, and to think "the less the slope the less the labour." Now I have noticed that when I am breasting a steep grassy mountain slope, I have a tendency to slew off to right or left, and thus to lighten my labour. And I thought I might make a few experiments with Toby (which I have subsequently repeated with Tony). I sent him down the slope again and again, after a small stone. At first, when he was fresh, he came straight up. But as he got tired he began to zigzag. He seemed to slew off in one direction or the other, just as I tend to do myself when I breast a steep slope. When he had slewed off a little way, he looked up at me and made straight for me, gradually slewing off in the other direction. So that the diagram of his course was thus—*a* being his starting-point, *b* the spot where I was sitting. There was nothing in the facts observed to negative the interpretation that the zigzagging was the outcome of sense-experience. The dog slewed off the direct course, because he practically found that progression thus was easier. And if the action is possible to sense-experience, we are bound by our canon so to interpret it, and not introduce a perception of the relation between angle of slope and labour involved.

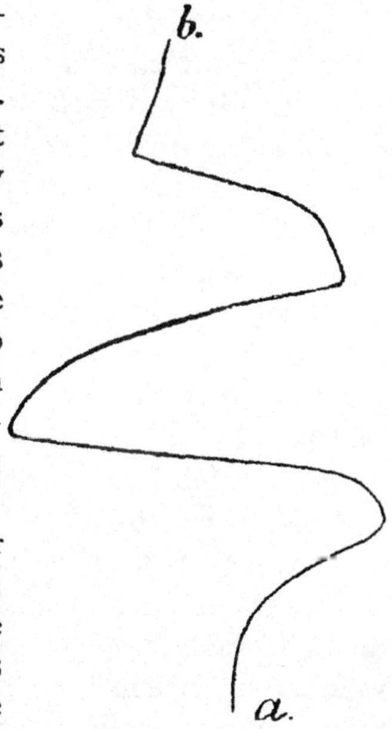

FIG. 20.

I must not bring this chapter to a conclusion without saying, that I freely admit that there are on record anecdotes of animals which I do not think can be readily interpreted as the outcome of sense-experience only,—instance the case contributed to Mr Romanes's collection by Mr Haden, of Dundee, of a monkey which wished to mount on to the roof of a hut by means of a door, and finding that the door repeatedly swung to, and thus baulked him, procured a thick blanket which lay in the cage and threw it over the door — the inference being that he perceived that the blanket would wedge the door open. That was a monkey worth experimenting with. If it really perceived this, and had no previous associations of casual origin on which to found its action, it could perceive other relations which experimental observation would have served to disclose. In the case of the *Cebus* monkey which Mr Romanes and his sister had for some time under close observation, there is nothing recorded that is not explicable as the outcome of intelligence; or so it appears to me.

On the whole, I am inclined to conclude that when we separate observed facts from observers' inference, there is a remarkably small percentage of cases in which the interpretation, on the assumption of sense-experience only, will not hold good. Such are the cases which should, wherever possible, be made the basis of an experimental investigation. As matters stand at present, I think it far more probable that the small percentage of outstanding cases would, on complete investigation, be shown to be the result of the exercise of intelligence, than that they involve reason in the sense in which I have used this word. I am very far from wishing to occupy the false position of dogmatic denial of rational powers to animals. I think it is a subject for further and fuller investigation. But I do express the opinion that the fuller and more careful the investigation, the less is the satisfactory evidence of processes of reasoning; and that, though the question is still an open one, the probabilities are that animals do not reason.

CHAPTER XVII.

SUBJECT AND OBJECT.

IT is not my intention to attempt in this work any detailed account of the steps by which the faculty of conception, applying its nascent power of analysis and synthesis to the data of sense-experience, impressional and transitional, may have gradually built up a scientific and philosophical interpretation of the world in which we live. The process has been a gradual one, involving several factors. How language would aid in this process I have already briefly indicated. If once, for example, the word "shining" or its equivalent is made symbolic of the conception of a particular quality exhibited by some individual visible object, it forthwith becomes also an instrument both of analysis and of synthesis. Of analysis, because it serves to detach the special quality from other qualities invariably associated with it in experience; of synthesis, because it serves as a centre of aggregation for similar qualities of other objects.

Such a grouping of objects in reference to particular qualities under the influence of the symbolic name, would form an initial stage in the reflective creation of an orderly universe out of the inchoate sequences of sense-experience. A further stage would be reached when the perception of these qualities became an end of conscious endeavour, and man looked out on the world, not merely as a witness, but as an observer. We are all, for example, witnesses of atmospheric changes from year's end to year's end; and perhaps some rough-and-ready generalizations are forced upon us, or are the direct outcome of our native wit. But we are content, for the most part, to remain mere witnesses. A few of us are, on the other hand, observers. These watch the changes of the weather, with the special object of perceiving the relations involved, and of rising to the conceptions of meteorology. When man thus becomes an observer, he takes an important stride towards the attainment of wider conceptions of the world. And when to observation he adds experiment, which may be described as observation under accurately controlled conditions, he renders his analysis more searching and extends the range of the synthesis dependent thereon. Most important, too, as an aid to the grasping of relations exceeding the reach of immediate perception, is the method of diagrammatic representation. The essence of this process is the translation of all relations, whatever their scope and nature, into visible space relations within the range of immediate perception. A map of a district or country thus condenses to within the reach of immediate perception space-relations of wide extent. An astronomical diagram enables us to grasp the relative sizes and distances of bodies, the actual sizes and distances of which tax the imagination to the utmost. The physicist can in this way represent the relative amplitudes of ether-vibrations of surpassing minuteness. Anything which can be expressed in numerical relations can thus be translated into perceivable space-relations, and thereby, through diagrammatic representation, brought home to the mind through the eye. It is well known that in this way we can represent the fluctuation of prices, death-rate, commercial prosperity and depression, statistics of crime, and numberless other widely different changes. I have used this method in all that I have said concerning the wave or curve of consciousness. Such a curve represents diagrammatically to the eye the relative intensities of numerous factors in consciousness. It, in common with other figures introduced into this book, is a mere diagram, and in no sense a picture. And it is essential, in the employment of this method of diagrammatic representation, that we should steadily bear in mind the fact that it *is* merely diagrammatic and symbolic or representative. More especially is this necessary in psychology, where we endeavour to make clear by physical or spatial analogies, changes in consciousness the nature and character of which are neither spatial nor physical.

Thus the progress of scientific interpretation involves as co-ordinate and concurrent processes,—first, inductive generalization through the continued application of analysis and synthesis; and, secondly, the testing of the generalization by further experiment and observation. The knowledge so gained is condensed in propositions, and brought within the range of perception through diagrammatic representation. Schematic logic does not afford much assistance in the process; it is an afterthought. The man of science, like the artist, is largely dependent on flashes of insight in moments when the wave of consciousness is peculiarly full, rich, and intense. How they come he may not be able to say, but he cannot often conscientiously attribute them to schematic logic. It is idle, therefore, to expect through the application of rules of scientific procedure to attain scientific insight; for the man of science in so far as he is creative is an artist. One can only say to him, as one would say to other artists:—Saturate yourself through and through with your subject, and with all that bears, or may bear, upon it, and *wait*. If the flash of insight comes, treasure it, and then patiently work it out in all its bearings, remembering that no art-product is made convincing without labour. *Then* you may apply your rules of scientific method with profit and advantage. And if it does not come, still *wait*, and meanwhile be content to serve science as a useful if uninspired day-labourer.

But I have said that it is not here my purpose to endeavour to trace the steps by which a scientific interpretation of nature has been, or may have been, reached. Such an attempt would be out of place in an Introduction to Comparative Psychology. There is, however, one aspect of such an interpretation, in its widest significance, to which we are bound to devote our careful attention. For the interpretation of nature involves the interpretation of consciousness as manifested in nature—in, at any rate, some animals, in other men, and in ourselves. We speak of mind and matter, of self and not-self, of subject and object. What are the relationships involved in these antithetical concepts? How have the concepts been reached? Does the conception of Evolution apply to mind? And if so, from what has mind been evolved? The student of human psychology can perhaps afford to leave these questions, or some of them, on one side. In an Introduction to Comparative Psychology we must attempt to deal with them even if the attempt merely serves to show how ignorant we are.

Taking first the relation of subject and object we shall find it most convenient to begin by considering the impression as a product of sense-experience. Let us suppose that a puppy has an impression of a bone. For the puppy there is neither object nor subject, there is merely the impression as a bit of real vivid experience. But *we*, who wish to explain the puppy's impression, submit it to analysis. And in the conceptual field of our thought it becomes, so to speak, polarized. At one pole there is the objective bone, and at the other pole the subjective consciousness of the puppy. We may then, in the light of this analysis, explain the impression "bone" in two ways. We may say:—There is an object in consciousness; or we may say:—There is a consciousness of the object. These are merely different modes of expressing the same fact of experience. But the former gives emphasis to the objective side of the impression, the latter to its subjective or conscious side.

There lies before me a crystal of quartz. The very language in which I state the fact implies the differentiation of the impression into object (quartz) and subject (me). This impression, as I look at the crystal, is just as real as anything can be. It matters not that the impression is set in a background of relations, and is thus raised to the level of a percept or a concept. The basal impression suggestive of these relations is there in the focus of my consciousness. But, *quâ* impression, it is neither subjective nor objective; it is both and neither. Both, inasmuch as it is the raw material which on analysis may yield the subject and the object; neither, inasmuch as, *quâ* impression, it is not analysed. And now suppose that we do submit it to analysis in our thought, and by abstraction reach the quartz as object in consciousness on the one hand, and the subjective consciousness thereof on the other hand. It is surely clear that on the score of such analysis we have no grounds for saying that either the quartz as object, or the subjective consciousness thereof, is capable of independent existence. The object and subject, involved in the sense-impression, are like the colour or the scent of a rose, *distinguishable in thought* but they are not *separable in experience.* We distinguish quite clearly the colour from the scent of the rose, but we know that they are inseparable in sense-experience. So we distinguish the objective and subjective aspects *of* the impression, but *in* the impression they are inseparable.

Let us now pass from the impression to the perception. Suppose that I perceive the similarity of the two inkstands on my writing-table. Here, again, we have that which we may describe in one of two ways. We may either say:—There is a relation in the focus of consciousness; or, There is a consciousness of the relation. We polarize the perception as we polarized the impression, and we distinguish the relation as an object of perception from the subjective perception of the object. So, too, with the conception. I have a conception of rotundity; and this I may describe in two ways. I may say:—There is an Idea in the focus of consciousness; or, There is a consciousness of the Idea. We polarize the conception as we polarized the impression and the perception. But here again we are only distinguishing in analysis and through abstraction two aspects of the conception which are in the moment of experience inseparable and indivisble. We are thus driven to the conclusion that throughout the whole range of experience from the most primitive sense-experience to the highest ranges of conceptual thought, subject and object are inseparable. There is no subject without an object. There is no object without a subject. We may distinguish the objective from the subjective aspects of our experience; but we have no grounds for regarding them as divisible or separable. We must not therefore picture to ourselves, or conceive the subject as something that is self-subsisting and independent; or the object as part of an external independent world, into which the subject may, or may not, be introduced; and the problem of psychology as the question how these two can be brought into relation. We must rather regard subject and object as inseparably united in experience; and the problem of psychology as the question how this two-faced unity has had its origin.

We are now in a position to complete the catalogue of the meanings of that most puzzling word "object," which we began in a previous chapter. We there distinguished, first, the object of sense-experience—this is the objective aspect of the impression; secondly, the object as percept, when the object of sense-experience is viewed in its perceived relations; and, thirdly, the object as concept, when it is viewed in its generalized relations. A snake, for example, may be an object of sense-experience to a secretary bird; it is an object as percept to the naturalist who is identifying it by comparison with other snakes; and it becomes an object as concept to the biologist who regards it is an example of an extensive group of animals. In all these cases the object of sense-experience is presentatively or representatively present to consciousness as a nucleus. It is focal to the eye of sense-experience; the difference is entirely marginal. It is a difference of background. But we must now extend the word object, first to that which is focal to the eye of perception, for example a particular relation; and then to that which is focal to the eye of conception, for example, relationship in general. In this sense an abstract and general Idea, such as virtue, or redness, or matter, may be an object in consciousness; not indeed an object of sense, but an object of thought. Thus there are five meanings to the word object—(1) the object of sense-experience, (2) the object as percept, (3) the object as concept, (4) the object of perception, and (5) the object of conception. All of them have this in common, that they are antithetical and complementary to the subject. But the fifth class, the objects of conception, fall into two categories, according as they deal with the objective or subjective aspect of experience. For both subject and object can become objects of thought. We must be careful therefore to grasp in what sense a writer uses the word object or objective, for it is a fruitful source of fallacies. He may mean by objective, that which is antithetical and complementary to subjective,—this is the wider signification; or he may mean by objective, all that pertains to the object of sense-experience,—and this is the narrower signification. Thus the explanation of the world in terms of matter and energy is an objective explanation in the narrower sense; but the subject, and its states of consciousness, though it may be an object of thought in the wider sense, can never be objective in the narrower sense. In further illustration of this point, let us take such an object as a crystal of quartz. It may be the starting-point of wide-reaching conceptions, according as it is dealt with by the psychologist on the one hand or the physicist on the other. These we may represent in tabular form thus :—

explained by psychology as the consciousness of the object, in terms of those subjective phenomena we name sensations, impressions, &c.	Quartz	explained by physical science as the object in consciousness, in terms of those objective phenomena we name matter and energy, &c.

Now, the conceptions, not only of physics but of psychology, are objects in thought; they are all objective in the wider sense; but only those on the right hand, or physical side of the above table, are objective in the more restricted signification of the word.

In sense-experience *as experienced*, there is neither subject nor object. There is just the impression which as before explained is neither or both. But in sense-experience *as explained by reflective thought*, the impression is polarized into object in consciousness and consciousness of the object; and I have asserted that these two though distinguishable are inseparable. It may be said, however, that nothing is easier than to separate the objective bone from the subjective puppy. Granted;—but then there is no impression. I merely assert that in the impression, as explained by reflection, the subjective aspect and the objective aspect are inseparable. If there be no impression I make no assertion about it. What I do assert is, that if the puppy either sees or remembers the bone,—if he has either an impression or an idea of it, then in that impression or idea there is the objective and the subjective aspect, distinguishable but inseparable. And if some one says, that surely the bone exists as such, whether the puppy or any other living being has an impression of it or not; I reply that I never thought of denying this; but that, as psychologist, it is for me to deal with impressions and ideas; and that at such times as the bone is not forming the objective aspect of an impression or idea, it nowise concerns me. While I look at it, the bone as the objective aspect of my impression is real; every bit as real as the subjective aspect of the impression; the reality of the objective and subjective aspects being strictly co-ordinate. And as psychologist I have nothing whatever to do with the bone, or the world of which it is a sample, except in so far as it forms the objective aspect of impressions or ideas, or the nucleus of *ideas* of relation or conceptual Ideas. It thus forms for me as psychologist part of the visible tangible objective world,—a world, I repeat, every bit as real as the subjective world of which it is the complement. With the world except as it presents itself in conscious experience we have here no concern.

All this may seem to some of my readers somewhat laboured and wearisome. And yet I think it is necessary, if we would attempt an answer to that most difficult question (so lightly answered by those who do not think deeply thereon) how far and in what sense the organism in the stage of mere sense-experience is conscious of its own existence, or of the sentience of other animals. Let us work down to this problem from above, passing from what we know concerning our own consciousness, to what we may infer concerning the consciousness of mere sense-experience. Man is self-conscious. He has reached a definite Idea of himself as subject. He conceives the Ego in marked antithesis to the non-ego. What then is this self which he so conceives? What is the Not-self, which he places in antithesis to it? In the light of what has already been said, the answers to these questions are not far to seek. The Not-self is the generalized concept of all that reflection has taught us concerning the objective aspect of the data of sense-experience; the Self is the generalized concept of all that reflection has taught us concerning the subjective aspect of our life experiences. Throughout the whole range of these experiences, the objective and subjective aspects, though distinguishable in reflection, are inseparable in actual experience. Hence we may say, that the Self and the Not-self are the generalized concepts which arise out of the distinguishable, but inseparable, aspects of experience.

But it is impossible to deal reflectively with the Not-self, and not to have forced upon our notice that there is therein something more than the objective aspect of impressions, and of the transitions between them. We seem forced to conclude that the transitions are due to an *activity* inherent in the not-self; and we seem forced to conclude that the activity is *orderly and determinate.* If there were no inherent activity, there would be no transitions; if the activity were not orderly and determinate, there would be no science or knowledge. So too, it is impossible to deal reflectively with the self, and not to have forced upon our attention the fact that we have something more than a sequence of states of consciousness. There is an activity which is selective and synthetic, which is orderly and determinate.

Now here we stand between two opposing schools of psychology. The one school contends, or at any rate is said by its critics of the opposing school to contend, that consciousness is a mere spectator or onlooker, curiously watching the play of material physiological forces, which it is incapable of controlling or modifying in the least degree. This, in condensed expression, is the doctrine of conscious automatism. On the other hand, an opposing school contends that the orderly selective and synthetic activity of consciousness is its essential feature, without which the subject tumbles to pieces as an incoherent series of sensations with nothing to give them unity, and to interpret them into a whole; and this school further contends, or at any rate is said by its critics of the opposing school to contend, that this orderly activity of the subject is something altogether apart from, and dissimilar to anything else in the whole realm of existence. Discussion and criticism of the views of opposing schools of thought forms no part of the scheme of this work, and I do but allude to the above-noticed divergence of opinion in order to mark more clearly my own position. Agreeing heartily with the view that there is an orderly and determinate activity in consciousness, in the absence of which the self would be inexplicable; and agreeing further in the view that this activity, *quâ* subjective, is distinguishable from any manifestation of activity, *quâ* objective; I believe that, so far from being dissimilar to anything else in the whole realm of existence, this selective synthetic activity in consciousness is but the subjective aspect of the selective synthetic activity which is objective in the not-self. If, as I said above, the self and the not-self are the generalized concepts we have of the distinguishable but inseparable aspects of experience, I do but add to this the conclusion, that an activity which is determinate, selective, and synthetic, is an inalienable part of this generalized conception. It is this inherent selective and synthetic activity, in its subjective aspect, to which the word *Will* * is properly applicable. And it naturally follows that those who deny or neglect the synthetic activity in consciousness, deny or neglect also the Will, and seek to resolve the phenomena into a sequence of presentations or representations. There is no place for the Will in a scheme of conscious automatism. That the Will may be omitted in psychology as descriptive, I admit; that it can be omitted in psychology as explanatory, I am not prepared to admit.

An illustration by analogy may here be helpful. If we

---

* It is well known that Schopenhauer applied it to the objective aspect also.

allow a concentrated solution of alum to evaporate, and hang therein a piece of thread, we shall see the formation of alum crystals of definite shape and orderly growth. Now suppose that such a crystal were endowed with reflective self-consciousness. It would perceive in the growth of its own or its neighbour's material "body," the movements of the molecules as they grouped themselves in crystalline form; for purposes of description (empirical) it would be sufficient to tell the whole story in terms of antecedence and sequence:—Such and such a disposition of the molecules in one moment is followed by such and such a disposition in the succeeding moment. So too in describing empirically the growth of its own "mind" (the subjective aspect of the molecular changes), it would be sufficient to tell the whole story in terms of antecedence and sequence:—Such and such a grouping of the elements of consciousness follows on such and such a preceding grouping. But if it were asked to explain how it was that these phenomena uniformly followed each other in this way, it might either say at once:—I do not know; or it might take one step, and a quite legitimate step, before confessing ignorance. It might say that a survey of all the facts empirically described, justifies the inference that they are the outcome of an activity which is synthetic, selective, and determinate. If it applied the term *will* to this activity in its subjective aspect it would further maintain that the will was *free*, just in so far as it was unhampered by external constraint. The building up of the crystalline structure and character to its natural development would be its typical example of freedom; but if many crystals were forming in a crowded space so that none could assume the natural form, this would be an example of external constraint.

So too do I, to apply the analogy, reflectively surveying my own bodily and mental development, see in both a sequence of phenomena due to an intrinsic activity. There are bodily or physical manifestations, and there are mental or psychical manifestations; and underlying both there is the intrinsic selective and synthetic activity. It is in virtue of this, that, in the language of philosophy, the individual is of common essence with the universal. And this assumes a new aspect when philosophy, touched by emotion to its finer issues, rises into religion. In virtue of the intrinsic activity I am I. And in virtue of the intrinsic activity the not-self, in so far as known to me, is what it is. The self and the not-self, I repeat, are the generalized concepts we have framed of the distinguishable but inseparable aspects of experience. Beyond experience and inferences based thereon, we do not go. But if a synthetic activity is the very essence of the self; and if the not-self, as known, is indeed its objective aspect; then is such a synthetic activity the very essence of objective nature. And thus we justify the conclusion that we reached at the close of the fifteenth chapter. Subject and object, no matter how completely we may distinguish them in analysis, are in essence one. And experience, alike in its objective and subjective aspects, is the determinate result of an activity which is selective and synthetic.

Self-consciousness would seem thus to involve—first, the conception of the subjective as distinguished from the objective; secondly, the concentration of the net result of all subjective experience into one generalized concept; and thirdly, the further conception of this net result as due to the determinate working of an activity which is synthetic and selective. This is self-consciousness in its most highly developed form. Not all of us, however, attain this degree of precision and clearness. For most of us the self is the imperfectly distinguished subject of our life's experience. And many of us would be unable to say whether the body is part of the self or of the not-self. Very vague is the schoolboy's or the peasant's conception of self, and for them there can be no doubt that the body forms part of the "me." Even for the most reflective among us there are periods—often periods of exceptionally healthy vigorous existence—when the concept of self never comes near the focus of consciousness. When I am playing a hard game of tennis, or when I am sailing a yacht close to the wind in a choppy sea, self does not at all tend to become focal. At these times the idea of self is in abeyance. Hence though I am a self-conscious being, I am not always self-conscious. And presumably at such times as I am least self-conscious, I am nearest the condition of the animal at the stage of mere sense-experience. What then does reflection tell me concerning my mental condition at such times? Although I am not occupied with an Idea of self, I am fully and vigorously conscious,—nay, rather I am exhilarated with the sense of pleasurable existence, my whole being tingles with sentient life, my wave of consciousness is exceptionally deep and full. In a word, I sense, or am aware of, my own life and consciousness, in an ususually vivid manner. But such sensing is marginal, and not focal. And just as I have before said, that in the life of sense-experience there is a sensing or awareness of relations not yet perceived or conceived (the word relations being used with anticipatory force), so now I would say, that in the life of sense-experience there is a sensing or awareness of the self that has not yet been distinguished in thought from the not-self. Experience is vivid and continuous. Such I take it to be the condition of the conscious but not yet self-conscious animal. And what, it may be asked, gives body and continuity to this sensing of existence? I have no doubt about the answer to this question. It is the fulness and continuity of the margin of subconsciousness, which forms a relatively constant setting to a sequence of varying impressions. As I stand by the helm of the yacht, the keen wind with an occasional dash of spray in my face, the swish of water along the counter, the pitching of the vessel; set teeth, firm lips, knitted brow, and muscles of arms, trunk, and legs well braced;—all this forms a relatively permanent background to the visual impressions, as I observe the advancing waves or watch the set of the mainsail to the wind. When we remember from how many thousands of nerves all over the body, not merely on the surface but also from internal parts, impulses are raining in upon the brain, numbers of which are no doubt infra-conscious, but many of which contribute to the subconscious margin of the field of consciousness; when we remember too how continuous is this stream of impulses, we shall not be at a loss to find the raw material of the sensed continuity of consciousness. Not that mere continuity, mere abundance of material, would serve to constitute even the consciousness of sense-experience. That consciousness, no less than the higher self-consciousness of man, is a product of selective synthesis ordering and grouping the material—as the molecules of a crystal are ordered and grouped, only with indefinitely greater subtlety and mobility.

Such then, it appears, may be the nature of that glad awareness of existence which an animal at the stage of well-developed sense-experience may have. Self-conscious it is not, and cannot be. For such an animal has not reached the level where perception and conception take their part in the mental development. And where there is no Idea of self, there is no self-consciousness properly so-called. But conscious they are, and they feel their existence to the full. It is only through reflection that experience is polarized into objective and subjective; and therefore, prior to reflection, the self as subject is not distinguished.

But if the animal at this stage, though sensing its existence in every moment of consciousness, has no conception of the self as the subject of experience, how does it regard other animals? Let us suppose that a puppy is at this stage; how does it regard its brother puppy? Were it capable of reasoning (which *ex hypothesi* it is not), it would be able to think thus:—When I am pleased I wag my tail; Spot is wagging his tail; therefore he is pleased. He would thus be able to regard Spot as conscious or sentient. But being incapable of such reasoning, is he able somehow to feel, if not to know, that Spot is sentient? I think we may fairly suppose that he is. Let us suppose that he yelps in a particular way when he is in pain. He hears the yelp that he himself makes, and there is associated with it the pain that he suffers. Now when he hears a similar yelp from Spot, there is called up by association a memory of the pain. His own yelp was set in a background of presentative pain; Spot's yelp calls up a background of representative pain. A hundred experiences of daily life would enforce a similar association. And thus the impressions which Spot produces in his consciousness would come to be surrounded with a fringe of sentience. It is possible that some one may be tempted to exclaim, "Ah! but this fringe of sentience is purely subjective; it is a revival of the pain the puppy itself has experienced, and is not associated with Spot as the object of sense-experience." But he who would thus exclaim must have forgotten what I endeavoured to establish with regard to the impression—that, as such, in naïve experience, it is not yet polarized into subjective and objective. It is simply a bit of direct unsophisticated experience. And in this simple direct experience, the impression "yelping puppy" carries with it a fringe of sentience.

It is no doubt difficult for man to divest himself of the relational halo which reflection adds to his simplest impressions—perhaps impossible to do so entirely. But if we endeavour, so far as possible, to catch ourselves in moments when sense-experience is most direct, and least modified by reflection, we shall, I think, find that it is the unanalysed impression itself that carries with it the fringe of pleasurable or painful sentience. We are sitting in the theatre, and the curtain rises on a group of merry village maidens. The impression is pleasurable. And the pleasure is not carefully assigned either to the objective young women or to the subjective ego. That is the business of reflection. It is simply that the living dancing-girls give rise to an impression of a particular order, and that this impression is suffused with pleasure. Turning the corner of a street, I once came upon two men supporting the senseless form of a poor fellow who had been knocked down by a cab, and was badly cut about the face. I well remember the faint thrill that ran through me as I shrank back. This, I feel sure, was the direct effect of the impression as such, and that it was, at the moment of experience, unanalysed into subject and object. Such, I take it, is the normal experience of the unreflective animal. And such, I may remark in passing, is the psychological basis and origin of sympathy.

We must remember how important a factor in animal life is imitation; how puppies, or kittens, or piglings of the same litter, act in concert. Not far from my school was a farm-yard, in which I watched many a litter of little pigs. Shall I confess that I sometimes visited the yard catapult in hand, and with this singled out a particular pigling? The shot was followed by a squeak and a rush, in which not only my particular quarry but the whole litter participated. Each pigling in the rush experienced certain feelings of alarm or fright, and at the same time saw his brothers running. These two, (1) the sight of another pigling scuttling off, and (2) a feeling of fright, would thus become associated. And in this way participation in common actions would beget a feeling of community in sentience. Where animals hunt in packs, as with wolves, this community of sentience would be proportionately strong. And then we must remember how wonderfully keen and acute are the senses of animals. The dog watches every change of his master's face, voice, and demeanour. Each has its associations as an impression.

Hence the extraordinary and beautiful sympathy of the dog. How largely sympathy depends on the appreciation of objective signs, we know among ourselves. The sympathetic man notices every varying expression and shadow of a shade. The unsympathetic person needs to have the fact that one is suffering or depressed shouted in his ear. The sympathy of the dog, so different from the behaviour of all but exceptional cats, is due to his keen receptivity and his social antecedents. The impression produced by every movement of his master is fringed with sentience. But there need be no reflective distribution of this sentience between subject and object. I am not denying that in the dog this reflective distribution may perhaps be superadded. But I submit that the evidence we have of sympathy in animals does not carry with it, as a necessary inference, that they have such powers of reflection and conceptual thought.

CHAPTER XVIII.

THE EVOLUTION OF CONSCIOUSNESS.

WE have, so far, taken for granted the existence of consciousness, and the fact that there are subjective phenomena which we, as comparative psychologists, may study. We have also proceeded throughout on the assumption that subjective phenomena admit of a natural interpretation, as the result of a process or processes of development or evolution, in just the same sense as objective phenomena admit of such interpretation. The question now arises: If consciousness exists, and if consciousness, as we know it, has been evolved, from what has it been evolved? It must be freely and frankly admitted that any suggestion which the comparative psychologist has to make, in attempting to answer such a question, must be speculative. But we must not be afraid of speculation so long as it is on scientific lines, and so long as the basis of speculation is honestly and fearlessly laid bare, and not slurred over and beclouded by ambiguous phraseology. Speculation is but the play of the imagination along the fringe which borders our knowledge; and imagination is the mother of insight.

It has before been pointed out that one of the cardinal steps in making our psychology comparative, and in linking this branch of science with those branches which deal with the objective aspects of our knowledge, is the correlation of psychical phenomena with physiological phenomena. It would be as idle as it would be disingenuous to pretend that this correlation had as yet been carried far. From the nature of the case, the difficulties of investigation are enormous. And much of the experimentation in physiological psychology — valuable as it undoubtedly is — deals with questions which are mainly physiological, and hardly touch psychology at all. But such work as has been done tends to strengthen and not to weaken the validity of the assumption; and, further, suggests that in and through this correlation we may seek a yet wider correlation of that which objectively we know as energy, and that which subjectively we know as consciousness.

Let us, however, put the matter in more concrete form. The dog that, as I write, looks up in my face with eyes so full of affection and of intelligence, suggests a double problem. How has that organic frame, with the bright brown eye and the warm active brain, been evolved, and from what? How has that keen intelligence and the consciousness it implies been evolved, and from what? We may look at the matter first and chiefly from the point of view of the development of the individual (ontogenesis).

In the common course of generation the dog is developed from a minute egg-cell, one hundredth of an inch or less in diameter, with which a yet more minute sperm has entered into fertile union. Supplied with shelter, warmth, and nutriment, by that maternal self-sacrifice which is a deeply significant fact of organic nature, this little speck of living stuff passes, by a process strictly continuous, though profoundly modified by the catastrophe of birth, into the dog, with its wealth of intelligence and affection. It is surely impossible, without extravagance, to speak of the fertilized ovum as conscious. Where, then, in the continuous process of development does consciousness come in? How, and whence? We are not now-a-days to be put off with the ambiguous assertion that consciousness and intelligence are "potentially" present in the germ. We ask: What is *actually* present therein as the basis of this potentiality? And if we are told that consciousness dawns at, or shortly after, the catastrophe of birth, then again we ask: Whence comes this dawning consciousness, and by what means does it become associated with the puppy's brain?

Having thus opened up these several questions, all of like implication, let us now endeavour to set forth the answer which seems most closely in accordance with scientific analogies. And to this end let us consider the living dog. His frame is pulsating with life and restless activity, and somehow associated with the transformations of energy in that brain of his there are states of consciousness. Were his skin, and the walls of his skull, as transparent as glass; did the molecular vibrations of his brain lie open to our curious scrutiny; could we trace in detail all the varied and orderly transformations of energy of which that brain is the theatre,—the accompanying consciousness would still be beyond our reach. *We* might follow the changes of energy; he alone would feel the states of consciousness. Now suppose that the dog dies. His body lies before us stiff with the *rigor mortis.* If we had weighed it previous to death, and if we were to weigh it again after death, the scales would give us no information of the departure of anything material. All signs of consciousness, however, are gone. And could we see through skin and skull into the brain, which during life was the theatre of so complex and orderly a sequence of transformations of energy, we should find that it was still and motionless. We are therefore justified in saying that, omitting minor qualifications, the orderly transformations of energy in the brain and the concomitant consciousness cease together at death. Closely associated during life, varying together in health and sickness, ceasing together at death, what is the nature of their connection?

Let us regard the matter from the objective aspect first, from the side to which the occurrences present themselves as transformations of energy. The state of consciousness being *ex hypothesi* accompanied by certain molecular vibrations in the brain or some part thereof, we have to note that from the physical point of view these molecular vibrations constitute an exceedingly complex and orderly mode of energy. It is upon this energy that we must fix our attention, the material structure of the brain being what we may call the vehicle of its manifestation. We are too apt to regard the *structure* as the essential thing on which to concentrate our mental gaze, partly, no doubt, because, through the invaluable labour of microscopists, we know so much that is definite about this structure. But a more penetrating insight enables us to see that the structure is merely the necessary basis of what is the really important thing—the manifestation of energy. The material structure of a steam-engine is of importance. But why? Because it is the vehicle for the performance of work. That is the really essential part of the business. In like manner nerve-structure is of importance. But why? Because it is the vehicle for the complex and orderly manifestation of energy. The essential importance of looking at the *going* machine, at the performance of work, at the energy of the matter in motion, not merely the material structure that is moved,—the essential importance, I say, of fixing our attention on this, being fairly grasped, we may now proceed to inquire from what the complex and orderly vibrations of the dog's brain have been evolved. In the fertilized ovum from which the dog was developed (and the same is true of the amœboid ancestor from which, hypothetically, the race of dogs has been evolved), there is certainly nothing approaching the orderly complexity of these molecular vibrations. But there are simpler organic modes of motion from which these complex molecular vibrations have arisen by a continuous process of development. It is from these simpler modes of energy in the simpler organic substance of the ovum that the more complex modes of energy which characterize the workings of the dog's brain have been evolved. In the development of the ovum into the embryo, and thence into the puppy and the dog, we may trace step by step all the stages of the evolution of those material structures which are the vehicles of these special manifestations of organic energy. We may watch the further and further differentiations of the nervous tissue, and the fashioning of the brain and its parts. It is true that we cannot indicate the exact moment when, in the increasing complexity of the tissues, the simpler forms of organic energy pass into the higher form of brain energy accompanied by consciousness. But that is just because it is a continuous development, an evolution. That the passage from the one into the other does actually take place we are bound, by all the canons of logical reasoning, to admit. It is only during life, however, that brain activity occurs, or is possible. A great number of modes of organic energy proceed side by side in the pulsating tissues of the living dog, their orderly continuance being what we term *life*. And only in and through their orderly continuance is the maintenance of the structure of the tissues rendered possible. The organic structure is like a spinning-top. Only so long as it spins and manifests its proper energy is its stability maintained. All around it are forces which tend to make it totter to its fall. But so long as it spins freely it can resist all minor attempts to upset its stability. And when the dog dies, what happens then? The specialized molecular vibrations of the brain, in common with other forms of organic energy, cease. The top no longer spins, and the structure totters to its fall. Decomposition sets in. The orderly organic changes which characterize life, give place to the destructive changes which characterize decay. But according to the law of the conservation of energy, although there is decomposition of the tissues of which the body was composed there is no destruction or annihilation of energy. The particular modes of energy through which the body was instinct with life pass away; but only to give rise to their equivalents in other modes of energy. Just as the puddle in the road disappears, but only to give origin to an equivalent mass of invisible water vapour; just as the candle disappears, but only to give rise to its equivalent mass in the products of combustion,—so throughout life and death the energy which throbs in the tissues neither appears nor disappears except at the expense of, or to the gain of, other modes of energy. Life is like a vortex in a rapid stream;—on surrounding energy it is dependent for its continued existence; into surrounding energy it melts away. And this is true not only of individual life, but of life in its entirety.

Turning now from the objective aspect to the subjective aspect, we pass from neural processes to states of consciousness. The states of consciousness in the dog's mind are the subjective aspect of what, from the objective aspect, are the molecular vibrations of his brain-tissues. And as in considering the matter objectively, so now, in regarding the mental aspect, we must ask from what the complex and orderly states of consciousness of the dog's mind have been evolved. In the fertilized ovum from which the dog is developed (and the same is true of the amœboid ancestor from which, hypothetically, the race of dogs has been evolved), nothing so complex as a state of consciousness is to be found. From what then have the states of consciousness been evolved? Do we not seem forced by parity of reasoning to answer:—From something more simple than consciousness, but of the same order of existence, which answers subjectively to the simpler organic energy of the fertilized ovum? Such, at any rate, is the hypothesis which appears to me the most logically consistent. It requires, however, no little effort of thought to conceive the existence of those elementary states from which consciousness may have had its origin. We may be aided in doing so, perhaps, if we fix our attention on the close association of brain-energy and states of consciousness, regarding them as *distinguishable*, but not *separable*. Now, the nervous energy of the brain is extraordinarily complex; and yet we believe that it arises by a process of continuous development from the much less complex energy of the fertilized ovum. In the ovum there is no brain-energy; there is only the far simpler germinal energy from which it is evolved. So, too, the consciousness in the dog's mind is wonderfully complex; but if it has arisen by a process of development, it must have been evolved from something of like nature, only indefinitely simpler. May we not fairly suppose, therefore, that in the fertilized ovum, though there is no consciousness, there are the germinal states from which consciousness may be evolved? Or, to put the matter tersely, may we not say:—As the complex molecular vibrations of the brain are to the simpler molecular vibrations of the ovum, so are the complex states of consciousness associated with the former to the simpler states of infra-consciousness, if we may so call them, associated with the latter. It is the association of consciousness and infra-consciousness with energy—its objective manifestation—that is the distinguishing feature of the view which I am endeavouring to set forth. Concomitant with the evolution of higher modes of organic energy from those lowly modes which alone obtain in the ovum or the amœba, is the evolution of consciousness from lowly modes of infra-consciousness.

We must now take a further step,—one, however, in which all evolutionists will not be prepared to follow us. For those, however, who believe that the organic has risen on this earth by process of natural development from the inorganic, the hypothesis must be more sweeping in its range. We must say that all modes of energy of whatever kind, whether organic or inorganic, have their conscious or infra-conscious aspect. Startling as this may sound, there is, I believe, no other logical conclusion possible for the evolutionist *pur sang*. For where are we to draw the line? The states of consciousness of the higher animals have been evolved from lower forms of infra-consciousness in the amœba-like or yet more simple protoplasmic germs in the dawn of life. But if those low forms of organic infra-consciousness were themselves evolved, from what could they arise, if they were not developed from yet more lowly forms of infra-consciousness, similar in kind, but inferior in degree, associated with inorganic modes of energy? In any case it is here submitted that this doctrine, that infra-consciousness is associated with *all* forms of energy, is necessarily implied in the phrase mental evolution for all thorough-going evolutionists who have grasped the distinction between consciousness and energy. And if this be admitted there is disclosed, by implication, an answer behind and beyond that ordinarily given to a question which has again and again been asked—the question: Is there a conservation of consciousness analogous to the conservation of energy? The negative answer generally given to this question results from the fact that the question itself has always been put in a form which does not admit of a satisfactory solution. There is not a conservation of consciousness any more than there is a conservation of nerve-energy or a conservation of electrical energy. There is no conservation of nerve-energy, because this is only one mode of energy which may be transformed into other modes. Not until we have generalized energy so as to include *all* its modes can we speak of conservation in reference to it. So, too, not until we have generalized that universal form of existence, of which consciousness is only the highest and most developed mode, so as to include all modes, can we speak of conservation in reference to it. But so generalized, I submit that there is a conservation of that form of existence which includes both consciousness and infra-consciousness, co-ordinate and co-extensive with the conservation of energy. Just as the dominant nerve-changes in the dog's brain are like a special vortex in the onward-flowing stream of the world's energy, so are the states of consciousness in his mind like a special vortex in the onward-flowing stream of that mode of existence which, whether it have risen to the level of consciousness or not, is still of the conscious and infra-conscious order. For the believer in scientific monism there is but one vortex, objectively presented as energy, subjectively felt in consciousness. For the dualist there are two vortices,—(1) an objective vortex, and (2) a subjective vortex, each associated with the other. In either case the vortex is dependent for its continual existence on surrounding stores of that out of which it has arisen; and in either case the modern tendencies of scientific thought suggest conservation, which is but the antithesis of creation *ex nihilo*.

A few words in conclusion, to present the matter from a somewhat different point of view. Consciousness exists: of that there is no doubt. How did it come to exist? There seem three possible answers to this question:—(1.) It was specially created in man, or in some lower organism from which man has been evolved; (2.) It has been directly evolved from energy; (3.) It has been evolved, as I have suggested, from infra-consciousness.

Now, the first answer—that of special creation—is, in my opinion, a logically tenable one, and one with which I have sincere sympathy. I do not hold it myself, because it does not seem to me either the highest or the most probable view of the matter; but if others hold it on these grounds, so let it be. With the second answer I am in distinct and direct antagonism. I do not think it has a single genuine fact of observation or a single rational inference from observation in its favour. Its supporters may be left to make out a case for it if they can. The third answer is that which I have endeavoured to set forth. If, then, these three answers exhaust the logical possibilities of the case, and if the second is inadmissible, through default of evidence in its favour, we are left in presence of the first and third. Either special creation, or evolution from infra-consciousness; there is no other alternative.

Accepting as I do the alternative of evolution, I nevertheless see in this evolution the continuous manifestation of a synthetic synthesis, which finds its expression in the primary laws of nature and of mind, and with which I shall deal further in the next chapter. Herein I seem to find the essence of the whole process, that which makes it comprehensible and rational. Regarding man physically and psychically as the crowning product of this evolution, I nevertheless conceive him to be the self-conscious outcome of an activity, selective and synthetic, which is neither energy nor consciousness; which has not been evolved, but through the action of which evolution has been rendered possible; which is neither subject nor object, but underlies and is common to both.

CHAPTER XIX.

SELECTIVE SYNTHESIS IN EVOLUTION.

I SAID at the end of the last chapter, that in evolution we may see the continuous manifestation of a selective synthesis, which finds its expression in the primary laws of nature and of mind. Allusion has more than once in the course of this work been made to this synthetic tendency in nature; and towards the close of the chapter on *Synthesis and Correlation* I said that consciousness is essentially a synthetic unity, and that herein we might recognize the subjective aspect of that selective synthesis which we may discern in diverse forms throughout the objective world of nature. I here propose to deal, so far as can be done within reasonable limits, with this selective synthesis. I shall endeavour to present a restatement of the fundamental traits which characterize evolution, and to indicate what I believe to be the essential solidarity of nature in all its modes, inorganic, organic, and conscious. It is, of course, true that the laws of inorganic development are not the same as the laws of organic development; and equally true that the study of mind introduces us to a new aspect of the developmental process. Notwithstanding these obvious differences, the evolution that sweeps through nature is, I believe, one and continuous. My chief concern here is with mental evolution, and it is towards this that I shall throughout be working. But I shall, nevertheless, devote a large share of my space to a consideration of the principles of evolution as they are exemplified first, under the simpler and more rigid conditions of inorganic nature, and then under the complex and plastic conditions of organic development.

If we make a nearly saturated solution of chloride of lead in hot water, and set the solution on one side to cool, we shall see after a while that myriads of minute acicular crystals of chloride of lead make their appearance and sink to the bottom of the vessel in which the solution is contained. Here is a simple case of development or evolution. Let us take note of some of the features it discloses. In the first place the crystals have a definite geometrical form, exhibit differential expansion under the influence of heat, and possess peculiar optical and electrical properties. In a word, the crystals are the result of a selective synthesis, special in its nature, and determinate in its products. Secondly, this selective synthesis can only manifest itself under appropriate environing conditions. If the water be hot, the crystals are not evolved; and if, after they have formed in a cooling solution, the vessel be again heated, the crystals will disappear. Thirdly, if we trace backward the evolution of one of these crystals, we reach a point, that at which the crystal began to form as such, where there is an apparent breach of continuity; by which I mean, not a gap or hiatus in the ascending line of development, but a point of new departure. Such new departures in development are illustrated in Fig. 21, which shows the volume-changes of the substance water under the uniform and continuous application of heat. It will be seen that the line which diagrammatically represents these changes exhibits several points of new departure. Opinions differ as to the exact state in which the chloride of lead exists in the solution, but, whatever that state, there is a critical period when the chloride of lead in solution assumes the form of a solid crystal. Fourthly, it goes without saying that the requisite materials out of which the crystal can be built must be present.

The laws of crystalline synthesis have to a large extent been ascertained, and it is found that all known crystalline

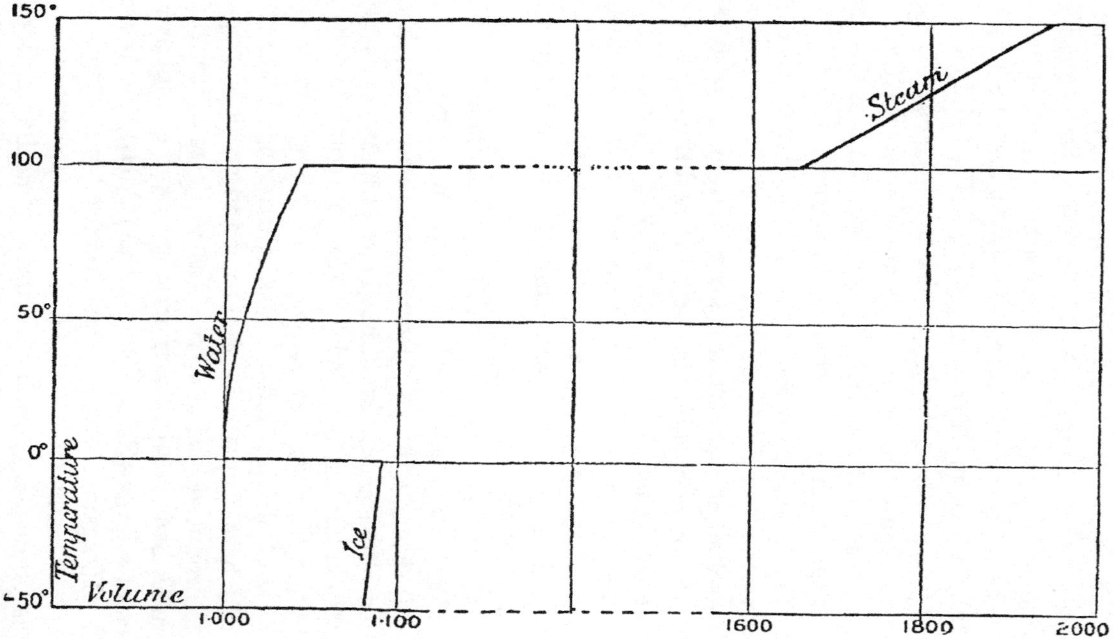

FIG. 21.

Diagram to illustrate breaches of continuity in development. It shows the effects of the uniform application of heat to the substance which assumes the forms of ice, water, and steam. The vertical scale represents temperature (Cent.); the horizontal scale volume. But on the right hand side of the diagram the scale of volume is different from (x 1,000) that on the left hand side: and the enormous expansion which occurs when water becomes steam is indicated by a broken line. Drawn for me by my colleague, Dr Sydney Young, F.R.S.

forms fall into one or other of six well-defined groups. There is a remarkable definiteness and, so to speak, narrowness in the limitation of crystalline form. Quartz crystallizes in six-sided prisms capped with hexagonal pyramids, the angles between the faces being of constant value. Though this fundamental form may be modified by the development of additional minor planes, the range of variation is slight. Calcite crystallizes in a greater variety of forms, and there can be little doubt that each one of these is the strictly determinate outcome of crystalline synthesis under the existing conditions of the environment. We must thus regard each several variety of crystalline form as the joint product of what we may term an innate or inherent synthesis and the conditioning of the environment. Were such a crystal endowed with consciousness, we may well suppose that it would claim freedom to act in accordance with its inherent crystalline nature, except in as far as it was restrained by the inevitable conditions of its environment.

The same general principles are further illustrated in that form of selective synthesis which is shown in the production of chemical compounds. If electric sparks be made to pass between carbon points in an atmosphere of hydrogen, acetylene, a gas containing two atoms of carbon combined with two of hydrogen (C_2H_2), is formed. If this gas be mixed with hydrogen and caused to pass over spongy platinum, ethene, which has two atoms of carbon and six of hydrogen (C_2H_6), is produced. If the acetylene be passed through a red-hot tube, benzene (C_6H_6) is formed, together with small quantities of styrene (C_8H_8), napthalene ($C_{10}H_8$), and retene ($C_{18}H_{18}$). If, again, electric sparks be caused to pass through a mixture of acetylene and nitrogen, hydrocyanic acid (HCN) is produced. Similar instances could be multiplied indefinitely. A mixture of carbon monoxide and hydrogen, when it is passed over spongy palladium, gives rise to formaldehyde (CH_2O), and six molecules of this in presence of calcium hydrate condense to form one molecule of a sugar called formose ($C_6H_{12}O_6$). Once more, ethylene (C_2H combines with bromine to form a substance called ethylene dibromide, which in presence of potassium cyanide in solution gives rise to ethylene dicyanide; this with water gives succinic acid ($C_4H_6O_4$), a substance found in the turpentine of several species of pine.

Now in each of these cases, and in thousands of other such cases with which chemistry makes us acquainted, we see that the new compound is due to a selective synthesis of the atoms, and that it has a definite and determinate composition. Not every collocation of atoms of carbon and hydrogen can coalesce to form a synthetic molecule; C_2H_2, C_2H_4, C_2H_6, are possible under the known laws of synthesis, but CH, CH_2, CH_3, CH_5, C_2H, C_2H_3, C_2H_5 and so forth, are, according to our present knowledge, impossible compounds. Secondly, this selective synthesis can only occur under suitable environing conditions. It is the function of experiment to ascertain the conditions under which synthesis can be effected. In the case of complex molecules, it is found to be quite impossible to produce them by a direct bringing together of the elements of which they are composed. They have to be built up by a series of successive steps or stages, the special conditions appropriate for each of these steps having to be ascertained by experiment. In this way many substances which are formed under organic conditions by animals or plants have now been produced in the laboratory, but many others cannot be so produced. We do not know the requisite stages, or have not learned the conditions under which the successive syntheses can occur. Here, again, we may notice, what we saw to be true of the development of crystals, that if we trace back the evolution of a chemical compound we reach a point, that at which the synthesis begins to take place, where there is an apparent breach of continuity, in the sense in which this phrase was before used. If the vapour of sulphur be passed over red-hot charcoal, a volatile compound, carbon disulphide, is produced, which may be condensed into a heavy, colourless liquid. This liquid has altogether new properties, different from those of either sulphur or carbon. It is a definite and determinate product of selective synthesis. Between the physical condition of the elements before synthesis and that of the compound after synthesis there is an apparent breach of continuity. There does not appear to be a gradual and insensible change from the physical properties of the elements to the physical properties of the compound, but at the critical moment of the constitution of the compound there seems to be a new departure.

Selective synthesis of a definitely determinate nature; the controlling conditions of the environment; apparent breaches of continuity in what we may term the curve of development: these, then, are some of the teachings of inorganic nature.

But it is possible that some of my readers are by this time beginning to exclaim:—What has all this to do with Comparative Psychology? I would ask them to bear with me. In the meanwhile, I may say somewhat in explanation of the central aim and purpose of this chapter. There are two opposing schools of psychologists. On the one hand are the Empiricists. They appear to be completely content with descriptive psychology. Laying great stress on the facts of association, they describe the genesis of psychological states as a continuous process of aggregation under the conditions of association. They do not appear to recognize—or in any case they lay little stress on—an underlying law of synthesis. Anxious to base their psychology on the results of biological research, and catching up from evolutionists the watchword "environment," they are apt to regard psychological genesis as wholly the result of the conditioning effects of the environment, and make the individual mind a mere puppet in the hands of circumstances. Allying themselves with the physiologists, they seem to regard consciousness as the mere spectator of a series of physiological changes in nerve-tissue. On the other hand are those whom we may term the Apperceptionists. For them the selective synthesis is the essential and central feature in mental development; the environmental circumstances are subsidiary. But they contend that this selective activity, to which they apply the term "apperception,"* is something *sui generis*, and peculiar to mind, something which is not found elsewhere in nature. "In the physical world," says Professor Mark Baldwin, "we find no such unifying force as that known in psychology as the activity of apperception." Now, according to the view adopted in this work, the Empiricists are right in their contention that the study of psychology should, so far as the conditions of the case admit, be brought in line with the study of branches of objective science; are right in utilizing to the full the results of general biology and of special nerve physiology; but are wrong in laying the main stress on the environment, to the entire, or almost entire, neglect of a law of intrinsic selective synthesis;—while the Apperceptionists, on the other hand, are right in insisting on the central importance of selective synthesis in psychology; are right in laying stress on the activity of the conscious subject; but are wrong in supposing that this selective synthesis is something *sui generis*, and peculiar to the activity of mind. Hence it becomes my duty to bring out, as clearly as I can, the real existence of this selective synthesis,—not as a mysterious "principle," but as a legitimate inference from the observed facts,—and also to show that this synthetic

* I cannot here discuss the use or uses of this term. Suffice it to say, that it appears sometimes to cover the synthetic tendency underlying all psychical processes, and sometimes to be restricted to certain higher and more specialized modes of synthesis, *e.g.* in the genesis of cognition or knowledge, and in the activity of attention.

activity is universal or common to all known aspects of nature, and nowise restricted to the realm of mind. And I do not see how I can hope to show this without a consideration of the phenomena of evolutional development in the inorganic sphere, as well as in biology and psychology.

Furthermore, I am anxious to draw a distinction which is, I think, too much lost sight of, but which is serviceable both in psychology, and in other branches of science,—namely, that between primary or intrinsic laws of nature, and secondary or extrinsic laws. Let it be once for all understood, that by law I mean the generalized statement of observed fact, or of legitimate inference from observed fact. The tendency of carbon to unite with sulphur is due to a primary or intrinsic law, and when they combine they are typically *free* so to do. And in all cases where natural inherent tendency is fulfilled without let or hindrance, we should speak of *freedom.* This is, indeed, the very essence of our idea of freedom, that the body or system of which this attribute is predicated is free to carry out the tendencies of its inherent nature. Unfortunately, false, and often hazy, conceptions have tended not only to obscure this truth, but even to lead to its denial. For if we conceive of the primary laws of nature as something external to phenomena and controlling them, then the sulphur and carbon are no longer free to combine, but do so under the compulsion of an external law. Regarding as I do the primary laws of nature as intrinsic and inherent (not external and compulsive), and any system exhibiting its own inherent activity as typically free, I distinguish these intrinsic laws from the extrinsic laws which summarize the conditioning effects of environment. Freedom is intrinsic; but all forms of constraint are extrinsic. The individual, man or animal or crystal, is free in so far as it acts in accordance with its inherent intrinsic nature; it is under constraint in so far as its activity is thwarted by the extrinsic influence of environing conditions.

These considerations should be borne in mind in any discussion of the free-will question. All natural law is determinate; hence there is no real conflict between determinism and freedom as above defined.

Now the selective synthesis which we have seen to be a factor in evolution is an intrinsic or primary law of nature; while the conditioning effects of the environment are secondary or extrinsic laws. Both are determinate, both are essentially natural. And it is my object to show that both are recognizable in evolution, in all its phases, inorganic, organic, and psychological.

I have spoken of the selective and synthetic tendency as *active*, and in the seventeenth chapter (p. 314) have laid some stress on this activity. That all the multitudinous natural operations we see around us, the changes and transformations of energy, give evidence of an activity of some sort and somewhere, can scarcely be called in question. But we are too apt to conceive this activity as external to and exercised upon natural objects; and to talk of the constraining action of natural law as if this action were something external and controlling. It is distinctive of the monistic interpretation of nature, that the activity is regarded as intrinsic or inherent in, and not external to, the happenings which we call natural.

In passing now from inorganic development to organic development, the first question that arises is:—Can the inorganic have given origin by natural processes to the organic? It seems tolerably certain that the synthesis of living protoplasm has not yet been effected in a test-tube; but he who bases upon this fact the assertion that the natural synthesis of protoplasm is impossible, would seem to have a somewhat overweening confidence in the test-tube. All that the failure to produce protoplasm in the laboratory shows, is that we have not yet learned, and perhaps may never learn, the stages of synthesis and the appropriate conditions under which the synthesis may take effect. I, for one, am not prepared to accept the verdict of the test-tube as finally conclusive. Whether the natural synthesis of protoplasm without the aid of already formed protoplasm is going on now anywhere in the wide world I do not know, but I believe that it has so taken place at some period or periods of the earth's history. The present existence of protoplasm shows that there is a synthetic tendency for certain widely-spread elements to combine in this way; nay more, the rapidity and vigour of organic growth show how exceedingly strong is the tendency. It is true that, so far as observation goes, this vigorous synthesis cannot now occur, unless a fragment of living protoplasm be present to initiate the new growth. But this does not nullify the fact of the existing synthetic tendency. Now one may well suppose that, notwithstanding the strength of this tendency, *the complete sequence of appropriate conditions for the successive stages of the synthesis is of the rarest occurrence;* may even have occurred only at a certain stage of the earth's history. If it be said that the properties of protoplasm indicate a new departure, and that there is apparently no continuity of curve between the properties of the inorganic and the organic, this may be readily admitted, and regarded as a marked instance of what we have already seen to be a not unusual feature of development.

Let us now take note of some of the more salient phenomena presented by living matter. It is in its chemical nature exceedingly complex, highly stored with energy, and very unstable. It is semi-fluid, viscid, and possesses mobility. It is capable of growth by assimilation, and of disintegration under the stress of comparatively slight stimuli or changes in the environment. It divides by fission into small individual units or cells, in each of which there is generally a central differentiated portion, the nucleus. There can be no question that this protoplasmic material, whether it be regarded as a single chemical substance, or as a group of analogous substances, is a product of selective synthesis; and there can be little doubt that the differentiation of the nucleus within the cell is a further result of selective synthesis, though how this was brought about we are not at present able to say. Furthermore, we may affirm with little fear of contradiction that, interesting and important as are the structure and composition of this living substance, it is chiefly in virtue of its being a vehicle for orderly transformations of energy that we describe it as living. Here, in fact, is where the organic differs most markedly from the inorganic. When the crystals of chloride of lead are formed in a cooling solution, there is a redistribution of energy; when the carbon and hydrogen combine to give rise to acetylene, or the carbon and sulphur to form carbon disulphide, changes in the distribution and mode of occurrence of energy occur. But in protoplasm, owing to the power of continuous assimilation and concomitant disintegration, there is a *continuous series* of transformations and redistributions of energy; and it is just herein that one of the characteristics of life lies. We study the substance of living things, not only to learn what it is, but also to understand what it does.

In the unicellular organisms, or those which are constituted by a single cell, there are wide differences in the complexity of the structure of this cell. Some, like the amœba, are comparatively simple; others, like the vorticella, show not a little differentiation; within the substance of yet others, exquisitely formed and delicately sculptured skeletons of silica or carbonate of lime are produced. That the more differentiated forms have been evolved from the less differentiated forms is part of the evolutionary assumption. But how? There can be little question that here, as on the lower plane of inorganic development, the two factors, an intrinsic synthetic tendency and a constraining and conditioning environment, are operative. It is difficult to assign the due value to each. When, however, we consider the exquisitely fashioned frustules of the diatom, the beautiful basket-work of the radiolarian tests, or the delicately chiseled calcareous shells of the foraminifera, we are unable to assign their conditioning to any probable mode of action of the environment. Nor have we any reason to suppose that the possession of those particular forms and modes of sculpturing which we observe, affords any security against elimination by the stress of inorganic conditions, by enemies, or by competition. We seem, therefore, justified in supposing that the primary factor is here the most important, and that these exquisite skeletons are the result of selective synthesis. If this be so, then the various types of frustules, to take the diatom as an example, must be regarded as determinate, synthetic products analogous to the crystalline forms displayed by calcite. But how different are the determinate forms displayed under organic conditions from those which are displayed under inorganic conditions!

There is no continuity between the crystalline forms of either silica or calcite, and the organic products of radiolarians or foraminifera.* But, from what we have already learned, such continuity is no necessary feature of development or evolution.

If now we pass from these skeletal products of certain unicellular organisms to the protoplasmic structure of such forms as vorticella, stentor, or paramœcium, with its not inconsiderable differentiation, and if we ask concerning this how far it is due to selective synthesis, and how far to the conditioning effect of the environment, our difficulties seem to increase. In either case the numerous forms that are, as a matter of fact, presented to our study must be due, on the hypothesis of evolution, to variation; and this variation must arise out of the inherent nature of the organic material

* Indications of continuity in so far as what we may term the ground-plan of structure, has recently been shown for radiolarians.

or some part of it, reacting under the stress of environing forces. The essential question, therefore, is whether the variations so arising are *indeterminate* or *determinate*. Are the forms we observe entirely the result of the natural selection of adaptive variations, through the elimination of an indefinite number of non-adaptive variations? or are they the result of the natural selection of adaptive variations from among those presented by determinate synthesis, numerous perhaps, but not indefinitely numerous? In other words, have we, when we reach this stage of evolution, got rid, for good and all, of determinate products of selective synthesis, and have we henceforth to deal with indeterminate variations in any or all directions? The question is a biological one, and most difficult to decide by observation. But, if the skeletons of certain unicellular organisms be regarded as the results of determinate variation, it would seem not unreasonable to assume that the differentiated structure of a vorticella or paramœcium is also the no-doubt naturally selected result of determinate synthesis. For it must be remembered, that the protoplasmic structure of vorticella, no less than the sculptured frustule of a diatom, is the visible expression of an intrinsic activity under the partial control of environing conditions.

The multicellular organisms differ from those which are unicellular, not merely and not chiefly in that they are composed of many cells, but in the fact that there are differentiations among the cells with differences of function; and that the differentiated structures and functions are so integrated and co-ordinated as to conspire to form a unity, both structurally and functionally. Among them reproduction is either by fission,—that is, the division of the organism into two or more, as in some worms; or by budding, as in many zoöphytes; or by the detachment of eggs which generally have to be, but in some cases need not be, fertilized. In sexual reproduction, to which alone we can here refer, the egg so detached is a single cell; but it is, so to speak, a representative cell. It contains the potentiality (to use this word for the present to express our comparative ignorance of the structure and energy in virtue of which the cell has this power),—it contains, I say, the potentiality of developing under appropriate conditions into an organism like that from which the cell was detached. This it is which makes the human ovum or sperm the most marvellous speck of matter in the known universe. Confining our attention therefore to the higher animals, we may say that each individual, with its tens of thousands of co-ordinated cells and cell-activities, takes its origin from a single cell, the nucleus of which contains germinal matter derived from two separate individuals. There can be no question therefore that within this cell, and probably within its nucleus, is contained the potentiality of the complex organism into which it develops. Here, then, it is that all variation must take its origin. In technical phrase, variation is germinal in origin, but somatic (from *soma*, the body) in expression.

The question here again arises how this variation in the germinal matter may be caused. It is clear that the extrinsic action of natural selection, through the elimination of relative failures, can have played but an indirect part in the origin of variations. Given favourable variations, natural selection may account for survival. It cannot account for their presence and origin. Variation is germinal in origin, but somatic in expression. It is on this somatic expression in the bodily organization that natural selection takes effect; and it is only in so far as these organisms are the bearers of germinal cells capable of reproducing similar individuals, that natural selection is effective on the race. Each individual carries, in the germinal matter, the secret of its structure and energy. If it perish through elimination, or if it be prevented from taking any share in the continuance of the race, its germinal secret perishes with it.

We seem, then, forced to conclude that the variations which take their origin in germinal matter must be intrinsic, and rather called forth by, than produced by, influences brought to bear upon it from without. If so, we must again ask, Are the variations determinate or indeterminate? According to one school of biologists, the variations are frequently determinate, and arise in a manner that is easily understood. If any part of the somatic structure of an organism is modified during the life of that organism, this "acquired character" affects the germinal matter in such a way that a similar modification is developed in the offspring which arise from that germinal matter. Thus, in so far as a man improves his physique or his brain-power by careful training, he is impressing his germinal matter with a potentiality of similar development. Those who hold this view have not as yet been able either to establish the fact of such inheritance of acquired characters beyond question, or to suggest such an explanation of the manner in which the influence affects the germinal matter, as to satisfy their critics, who contend that no such origin of determinate variations has been proved or is probable. According to the opposing view, all variations arise endogenously within the germ-plasm; and there is no transference to the germ-plasm of exogenous variations taking their origin in somatic modifications. If this be so, we are thrown back upon such influences, acting on the germinal nuclear matter, as nutrition, climate, and the nature of the medium in which the organism lives, in so far as this can affect the germinal matter in its sheltered position within the body, and upon the general effect of the organic environment of the somatic cells and their products. These influences may give rise to indeterminate variations, or to variations which are determinate. In the former case, we may say that chance offers an indefinite number of fortuitous variations to the winnowing process of natural selection. In the latter case, we may say that selective synthesis offers a definite number of determinate variations to the eliminating agencies of life's struggle. The observations of Schmankewitsch appear to show, that by gradually altering the salinity of the water in which certain brine-shrimps live, one species can be transformed into another species differing *determinately* in the form of the tail-lobes and the character of the spines they bear. Pupæ of a Texan species of the *Saturnia* were brought in 1870 to Switzerland and kept there during the winter. The moths which emerged from the cocoons were completely Texan in character, and laid eggs; the caterpillars were fed on the leaves of *Juglans regia* (the Texan form feeding on *Juglans nigra*); and the moths which resulted from this development were possessed of new and *determinate* characters such as to justify, it would seem, their being regarded as different species. Now here it is difficult to say whether the difference was called forth by some action on the germinal matter or on the cells which developed therefrom. But, in either case, the apparently determinate nature of the change is to be noted. It appears to me that the recorded experience of breeders shows that the material offered to artificial selection is not the result of fortuitous but rather of determinate variation. As Darwin said, "There are two factors: namely, the nature of the organism, and the nature of the conditions. The former seems to be much the more important." I am therefore disposed to believe that the factor that I have termed selective synthesis is still operative in the case of the multicellular organism. It is, no doubt, indefinitely more complex than it is in the case of inorganic development, or even of the development of the exquisite markings on the frustules of diatoms. But this increasing complexity is just what a study of the lower phases of development would lead one to expect.

I have spoken of the nuclear matter of the germ as containing the potentiality of developing into an organism like the parent. Dr Weismann has recently made an attempt to suggest a structural basis of this potentiality. The attempt is from the nature of the case speculative, but the speculation runs on scientific lines. I cannot, of course, here discuss or even describe Dr Weismann's views. Suffice it to say, that he suggests the possible or probable existence of biological units of progressive complexity, which he terms biophors, determinants, and ids. These give actuality in the germ to its potentialities of development; and they may well be regarded, from our standpoint, as the structural expression of determinate selective synthesis. Dr Weismann is of opinion that the cause of hereditary variation must be due to the direct effect of external influences on the biophors and determinants. "We are," he says, "undoubtedly justified in attributing the cause of variation to the influence of changed external surroundings." It is not, however, quite clear whether he regards the variations so produced as determinate in direction or not. But he forcibly states his belief that we cannot possibly attribute the immense number of adaptations to rare fortuitous variations occurring only once. "The necessary variations from which transformations arise by means of selection must," he says, "in all cases be exhibited over and over again by many individuals." If the variations are not indeterminate, but the outcome of selective synthesis, the occurrence of such variations as are exhibited over and over again by many individuals would be what we should naturally expect.

Granting all Dr Weismann's assumptions, the manner in which the determinants and biophors are distributed throughout the organism during the process of development remains an unsolved riddle. Of this Dr Weismann himself is fully aware. He is far too honest an inquirer into truth to attempt to gloss over our present ignorance.

It seems to me that the problem of development will have to be attacked in the direction rather of energy than of structure. We must regard the living organism, nay even the fertilized egg-cell, not only as a piece of mechanism, but also as a piece of going mechanism. I remember once seeing a lecturer build up a number of gyroscope tops into a compound system of spinning parts. So long as they continued spinning the system was stable; as the spinning died down the system fell to pieces. It is the spin of life that somehow effects the distribution of structure; and different eggs develop into different animals in virtue of an inherited difference of their vital spin. Cell-division cannot be a matter of structure only; it must be a matter of what I am speaking of as spin—a term which must not be interpreted too literally. When the organism has been built up into its several parts, composed of differentiated cells, the spin of each cell must in large degree be dependent on, and conditioned by, the spin of neighbouring cells which constitute its immediate environment, and must in turn influence that of those cells; and thus the spin of the entire group is a co-ordinated and integrated spin, conditioned as a whole by the general environment of the aggregate.

What I before termed the potentiality of the fertilized egg-cell, we may now regard as a definite biophoral or other structure, which is the vehicle of a particular and specialized vital spin. I am inclined to regard the vital spin as that upon which the more stress will ultimately be laid; the spin determining the development of structure, rather than the structure determining the nature of the spin. Starting, that is to say, with a particular and specialized spin in the fertilized egg, then, given an adequate supply of material fitted for assimilation, the evolution of the indefinitely more complex and co-ordinated spin of the differentiated cells of the adult organism follows, and is manifested to our eye as a structural product, the functional activities of the organism being the net result of specialized co-ordinations of vital spin. The analogy of mechanics may mislead us. We make machines with a co-ordinated mechanism to effect certain transformations of energy, and thus regard the structure as determining the nature of the output of energy. But in nature the machines are not made, but develop, and their structures are determined by the modes of energy of which they are the vehicle. On the view of evolution developed in this chapter the transformations of energy involved in the complex vital spin are due, primarily, to inherent and determinate selective synthesis of the vital molecular vibrations which constitute the spin, and secondarily, to the conditioning of the spin in harmony with the environment.

The vital molecular movements and transformations of energy reach their acme of delicate co-ordination in the brain of the higher vertebrates; and there they are associated with the phenomena of consciousness. We thus pass from organic evolution to mental evolution.

The essential questions for our consideration are: (1.) Is there selective synthesis in mental development? (2.) What is the nature of the conditioning environment in mental evolution? (3.) Is variation determinate or indeterminate? (4.) Are there apparent breaches of continuity in mental development? (5) Is mental development dependent on natural selection through elimination? (6.) Is it necessary that we should believe in the inheritance of acquired characters?

In reply to the first of these queries, Is there selective synthesis in mental development? it appears to me that the psychologist is bound to answer that there is. And I would remind the reader of what was said concerning the externality of the object as presented in vision. I look up from my page and have an impression of a book at a certain distance from me. In psychological analysis it is found that the impression is the result of the coalescence of certain retinal sensations with certain motor sensations derived from the eye muscles and the apparatus of accommodation. It appears to me obvious that no mere commingling of disparate sensations could give the out-thereness of the impression, but that when the motor sensations coalesce with the retinal sensations they enter into a synthesis which has a new and determinate character. It would seem that just as the raw material of life is the product of selective synthesis, so too is the raw material of sentience dependent upon an analogous process; and that it is inconceivable that the elements of sentience should give rise to consentience and consciousness by mere fortuitous grouping, without the play of that selective synthesis which sweeps, as I believe, through the whole ascending curve of evolution, inorganic, organic, and psychical. In any case, psychology, so far as I am able to interpret its teaching, proclaims the fact that selective synthesis is of the very essence of mental development.

Passing to our second question, we have seen in the lower phases of inorganic and organic evolution how important a part the environment plays in conditioning development. The products of selective synthesis must be in harmony with the environment, if they are to exist and persist. What, then, is the environment in mental development? We must here distinguish between the environment of the co-ordinated system of ideas which constitutes the complex mental synthesis of the individual mind, and the environment of the several units which are the elements in this synthesis. The mental synthesis is the correlate of the integrated spin of a vast number of brain-molecules; the results of this integrated spin must be in harmony with the general environment; but the spin of any subordinate group of cells must also be in harmony with the spin of the surrounding groups which constitute its immediate environment. So the mental system of the individual must be in sufficient harmony with the surrounding social and other environment to enable the man in whom it occurs to escape elimination by competition or by enemies, and to elude temporary elimination in the prison or the madhouse. On the higher plane of intellectual thought it must, in order to be fruitful and not barren, be fitted to resist the elimination of adverse criticism. Within the system itself the environment of any constituent idea or ideal is wholly psychical. This is obvious enough in the higher region of intellectual ideas, and of moral and æsthetic ideals. But it is also true in the domain of sense. For there is no test beyond and behind that of practical experience; and practical experience is psychical. Hence it follows, that any product of mental synthesis must, in order to make good its existence, be in congruity with the mental system in which it develops, Its psychical spin is conditioned by the co-ordinated psychical spin of the whole system. I am thinking out a problem and a solution flashes across my mind through some subtle association by similarity. But it has to stand the test of the environment. If it be not in congruity with all I know of the subject it cannot find a permanent place in my system of ideas: it is eliminated through incongruity. Contradictories cannot co-exist in the same synthesis. The advance of science is on this wise. The thinker assimilates all that is best in the work of his precursors, and by observation and experiment brings in fresh stores of facts; a congruous system takes form in his mind, all that is incongruous therewith being eliminated; and for after thinkers the result of his thought is part of the environment which they must assimilate, and which through the rigorous application of scientific method and verification they must hand on in a more highly developed condition through the further elimination of incongruities.

Whether the variations that occur in mental faculty are determinate or indeterminate in their direction (our third question), it is hard to decide. Are there definite types of character? or are what we regard as such types merely convenient categories under which we class individuals who are indefinitely variable? Is genius the result of a fortunate commingling of inherited aptitudes, or is it a product of selective synthesis, a crystalline gem of rarest purity and symmetry? I am inclined to believe that the variations are determinate, the definite products of selective synthesis, and that mental evolution proceeds along lines which are determined by the intrinsic laws of mind, just as a crystal is evolved along lines which are determined by the intrinsic laws of crystallization. But there can be little doubt that, as we advance from the simpler to the complex, the rigidity and narrowness of the synthesis give place to far greater plasticity and freedom. The determinate possibilities of synthesis are in brain and mind indefinitely increased, and thus, from the very multiplicity of determinate variations, their definitely synthetic nature is liable to escape our observation.

Passing now to our fourth question, Are there apparent breaches of continuity in mental development? I am disposed to answer that such apparent breaches there are. The step from mere sentience to consentience probably involved such a breach or new departure in the developmental curve. The step from consentience, or sense-experience, to reflection and thought certainly involves, in my judgment, such a new departure. The curve of the development of sense-experience and intelligence pursues a smooth upward course. But when the perception of relations is introduced there is a point of new departure. The curve takes a wholly new direction, and sweeps up to the highest products of rational thought. This is diagrammatically represented in Fig. 22, in which the changed direction of the curve is intended roughly to represent the introduction of reflection. If the dividing line between sense-experience and reflection is to be drawn between the lower animals and man, then we may say that there is a breach of continuity of development at this stage of evolution analogous to the breach of continuity between the

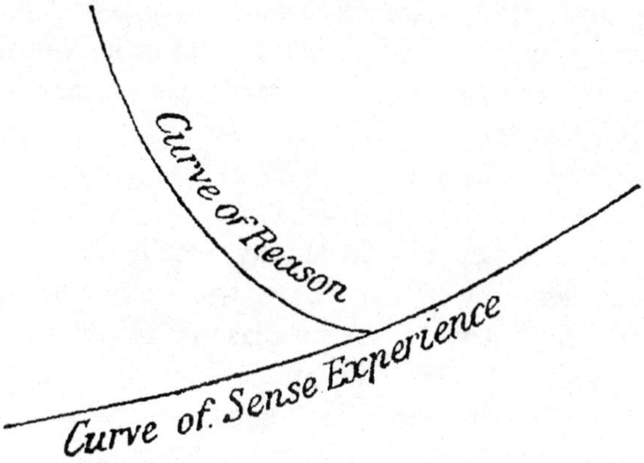

FIG. 22.

inorganic and organic phases of development. This is the view to which I myself incline. But we have seen that such breaches of continuity are natural incidents in the ascending curve of evolution. If, therefore, they do thus occur in mental evolution, *this does but bring this phase of development into line with the inorganic and organic phases.*

To the question, Is mental development in all its phases entirely, or even mainly, dependent on natural selection through elimination? I reply with an emphatic *no.* I do not of course deny that in the animal world, and in human society to a less degree, it has been the function of intelligence and reason to enable the organism so to guide its actions as to resist elimination, to live out its full span of life and to procreate its kind; and that those organisms in which intelligent adaptation was inadequate to these purposes have again and again suffered the penalty of elimination, leaving others fitter than them in possession of life's field.

But I see no evidence to show that commanding intellect, mathematical or scientific ability, artistic genius or lofty moral ideas, are attributable solely to natural selection through elimination.* In his essay upon "The Musical Sense in Animals and Man," Dr Weismann says:—"Talents for music, art, poetry, and mathematics do not contribute towards the preservation of the human species, and therefore cannot have arisen by the operation of natural selection." He suggests that as the dexterity of the hand was evolved for other purposes than piano-playing, but has been utilized by man for this purpose, so faculty, which was evolved for other purposes may by man have been devoted to mathematics, art, or poetry. I see no reason for disagreeing with this contention; but at the same time I submit that the development of definite and self-consistent artistic, musical, poetic, and ethical ideals demands an explanation which is nowhere given in Dr Weismann's essay. Such explanation is afforded, if they be regarded as the results of inherent activities which are selective and synthetic in their nature.

Lastly we come to the question, Is it necessary to our conception of mental development that we should believe in the inheritance of acquired characters? Much of the essay of Dr Weismann, to which I have just alluded, is devoted to the consideration of this question with special reference to musical faculty. He concludes, that such inheritance of acquired faculty is neither proved nor necessary to an adequate interpretation of the facts. In this I am disposed to agree with him; though I am not prepared at present to assert that such inheritance is impossible. The vital spin in the germinal cells may possibly be in some way influenced by the somatic spin around them; or the spin of the

* Mr Wallaschek has recently shown the practical value of primitive music among savages. And no doubt *the beginning and early stages* of our moral, intellectual, and æsthetic faculties, were of practical value in the struggle for existence.

biophors derived from specialized determinants may influence in an analogous manner the spin of the similar determinants in the germinal cells. Be this as it may, it appears to me, that if, as I have above contended, the development of definite and self-consistent artistic, ethical, and other ideals is due to selective synthesis, under the conditioning restraints of a psychical environment, we have herein all that can reasonably be required by any one who is content to adopt an interpretation of nature, including psychical nature, based on the principles of evolution. An activity which is selective and synthetic is disclosed throughout all the operations of nature, and in psychology is an essential factor in mental development. But there is no evidence that this activity is peculiar to psychology, and there is no evidence that it is external to, and not naturally inherent in, the phenomena which it is the business of empirical psychology to describe.

CHAPTER XX.

THE PSYCHOLOGY OF MAN AND THE HIGHER ANIMALS COMPARED.

IN the last two chapters I have endeavoured to indicate the relation of mental evolution to evolution in general, and to show that the selective synthesis which gives unity to the individual mind is of like nature with that which a study of evolution discloses throughout natural occurrences.

We must now take up the subject where we left it at the close of the sixteenth chapter, in which I contended that a very large percentage of the activities of animals may be fairly explained as due to intelligent adaptation through association founded on sense-experience. I freely admit that there is a small—in my opinion very small—outstanding percentage of cases, the explanation of which seems to involve the attribution to animals of powers of perception and of rational thought. But seeing the smallness of the number of cases of this type, and seeing the anecdotal character of the record, it is my opinion—an opinion which I shall have no hesitation in changing, if the results of systematic investigation and carefully conducted experimental observations warrant my so doing—that, were all the circumstances known, this outstanding percentage would disappear, and that the whole range of animal activities would be explicable as the result of intelligent adaptation. If this be so, then, in comparing the psychology of man and the higher animals, the radical difference lies in the fact that man perceives particular relations among phenomena, and builds the generalized results of these perceptions into the fabric of his conceptual thought; while animals do not perceive the relations, and have no conceptual thought, nor any knowledge—if we use this word to denote the result of such conceptual thought. Whether this conclusion (or hypothesis, if the word be preferred) is valid or not, will have to be settled, if it can be settled at all, not by any number of anecdotes,—interesting, and to some extent valuable, as such anecdotes are,—but by carefully conducted experimental observations, carried out as far as possible under nicely controlled conditions.

It is not my intention here to go over again the ground we have already covered. Enough has been said on that aspect of the comparative psychology of men and animals. But there are other aspects, on which little has been said. The emotional aspect of the psychical life has, for example, received but little consideration.

In the chapter on *Automatism and Control*, I said that primitively, and in the lower organisms, control is determined by the predominance of pleasurable or painful tone in the sensory centres which are at any time conspiring to influence the centre of control. But I added, that for man, in so far as he is a reflective being who frames ideals of conduct, this statement was too crude, and was contradicted by experience, unless we extend the meaning of the words "pleasurable" and "painful" in a way that can scarcely be regarded as satisfactory.

Now, the emotional states of men and animals are extraordinarily complex. Associated with every sensation-element there is, or may be, an emotional tone. I shall use this expression to denote the aspect of the sensation in virtue of which the organism which experiences it tends either on the one hand to seek its continuance or its repetition, or on the other hand to exclude it from consciousness. Now, as we have already seen, the states of consciousness which constitute the psychical wave of empirical psychology are exceedingly complex, with focal impressions or ideas and a marginal setting. Motor sensations contribute very largely to the states of consciousness. And what we call emotional states are the net result of the summation of emotional tone of all the sensation-elements which in varying degrees contribute towards states of consciousness. No wonder that their complexity is such as to baffle analysis. Moreover that synthesis which we have seen to be so important in sense-experience, and in the perceptual and conceptual superstructure that is founded thereon, is not less important in the emotional aspect of conscious experience. It is practically impossible to analyse even such a relatively simple emotional state as anger into its constituent elements of emotional tone. There is little doubt that the predominant impulses, with the effects of which the emotion is associated, are motor impulses. But the synthesis of these is carried out below the threshold of consciousness; their dissociation point, to borrow again an analogy from chemical science, is in the infra-conscious region. For psychology, as such, they are undecomposable. Just as we are consciously aware of only the net results of a great number of motor impulses, which synthetically combine to give rise to our motor sensations,—so too are we only consciously aware of the net results of the vast number of concurrent and conspiring impulses whose emotional tones synthetically combine to constitute that complex product which is felt as diffused throughout the whole body, focal and marginal alike, of the psychical wave in the moment of experience, and to which we apply the phrase "an emotional state." I say focal and marginal alike; for it would seem like omitting the central character of the drama, if we excluded the focal impression or idea, to the presence of which the emotion owes its origin. I am, however, disposed to regard the emotional state itself as mainly a matter of the marginal background of consciousness. Yonder fox-terrier who has caught sight of his old enemy the butcher's cur is brim-full of emotional tone all down his ruffled back to the very tip of his tail. The cur is in the focus of his consciousness; but it is set in a background of emotion that is thrilling in from every fibre of his frame. So too the mongrel that limply cowers in abject fear has the stick of his ruffianly master in focus; but it is the margin of consciousness that is trembling with emotional tone. The more I study the emotions the more do I feel convinced that they are marginal to consciousness, a matter of the mental background. And this fact serves further to increase the difficulty of any adequate analysis and classification of the emotions.

I do not propose to attempt here any detailed consideration of the emotional aspect of the practical life of sense-experience. I think that comparative psychology may fairly assume that throughout the range of the sense-experience, common to men and animals, their emotional states are of like nature with ours. And wherever the activities prompted by sense-experience have reference not only to the individual performer, but to other organisms, for whom, with whom, or against whom they are carried out, the associated emotional states, which, it must be remembered, constitute only the emotional aspects of sense-experience, have not only an individual, but a sympathetic bearing. The sympathy is indeed in merely the sense-experiential, not the reflective and self-conscious, stage. But it forms the basis of that higher sympathy which differentiates the social life of man from the social life of animals.

What we have especially to note is that the perception of relations, and the conceptual thought which grows thereout, brings with it a new order of emotional elements—those emotional tones which are associated with the relations themselves, which are synthetically woven into the already complex web and woof of the emotions of sense-experience. It is difficult to disentangle these threads and separate them out as the results of psychological analysis. They are associated with relations; and in relation they have their true value. But their introduction modifies in marked ways the emotions of sense-experience with which they are subtly interwoven.

Let us take as an example the pleasureable emotion that is aroused in connection with the so-called sense of beauty. I do not think that anyone who knows how the bower-bird decks its home, collecting flowers and fruits of bright and varied colours, removing everything unsightly, and strewing the ground with delicate moss; or how the humming-birds decorate their nest—"with the utmost taste," as Dr Gould observes—weaving into their structure beautiful pieces of flat lichen;—I say that I do not think anyone who knows his facts, can deny that some animals have a sense of beauty, and derive pleasure from objects which to them and to us are delightful to the eye. This is the outcome of, and is in close association with the life of sense-experience. But when the delicate hues and proportions of the objects are not only sensed, but perceived in relation to each other, then to the mere pleasure of sense there is added the higher æsthetic pleasure due to the emotional tone associated with the perception of relations. The Somersetshire rustic stands unmoved before the delicate geometrical tracery in the windows of the chapter-house of Wells Cathedral. It is not an object which appeals strongly to the *naïve* eye of sense-experience; the relations of the parts must be perceived before the beauty of the window can be appreciated at its true worth. It is the emotional tone which is associated with the perception of relations that is woven into the texture of the emotions of sense-experience, and that gives them a new value. This it is which raises the merely sensuous emotional tone of sense-experience to the æsthetic tone of the perceptual and conceptual phases of mental development.

There can be little doubt that the song of the nightingale gives pleasure to the singer, and we may fairly presume that it gives pleasure to his mate. For many of us the pleasure derived from music is of like order, and is to a large extent due to the emotional aspect of a specialized form of sense-experience. But when the transitions in consciousness are not merely sensed, but are perceived, and the subtle relationships of the melody and the harmony are appreciated, then there are added to the *naïve* pleasure of sense-experience new elements which raise its quality, enhance its value, and render it æsthetic. And this enhancement, it should be noted, is not due to the perceptual or intellectual element *qua* intellectual, but to the intellectual element in its emotional aspect. If the intellectual element as such, that is to say on its cognitive side, is obtrusive, it does not add to, but rather detracts from, the value of the emotional state. To get the highest emotional value out of music or any other art-product, we must abandon ourselves to the luxury of enjoyment in a wise passivity of intellectual self-surrender. But the intellectual elements, in their emotional aspects are there, and swell the volume of the pleasureable emotion. And if animals are limited to sense-experience, then in their emotional states, elements due to the emotional tones of perception are necessarily absent.

Now, when once the power of perceiving relations is introduced into the emotional aspects of experience, it must lead up sooner or later to a perception of the relations of the emotional states themselves to each other. In this way æsthetic judgments have their birth and origin. But there are two phases or types of æsthetic judgment; and since it might be contended that animals, even if they do not attain to the one, may at least reach the other, it will be well briefly to consider them.

In the first place we may notice that just as the perception of relations derives its raw material from sense-experience, wherein the transitions in consciousness, with which perception deals, are already sensed, so does the æsthetic judgment find its raw material in a sense of satisfaction or dissatisfaction, which arises according as an experience is pleasureable or the reverse; or, to put it on a broader and more satisfactory ground, as it is congruous or incongruous with the psychical nature. That experience which is congruous with the psychical nature is an object of appetence; that which is incongruous is an object of aversion; that which is merely not congruous, is simply ignored. When the animals—if I may illustrate my meaning by a mythological example—heard the strains of Orpheus's lute, there fell upon their ears sounds new to their experience; and since these sounds were pleasureable, or congruous with the psychical nature, they were objects of appetence, and the beasts drew near to listen; but if for Orpheus there had been suddenly substituted a beginner on the violin, feeling his way through a difficult and unfamiliar exercise, one may take it that the sounds would be objects of aversion, and the animals would have scattered far and wide. The sense of satisfaction, or the reverse, would in the animals be natural incidents in the life of sense-experience, and would imply no judgment. A felt congruity or incongruity, pleasureable or painful, gives only the germinal matter of æsthetic judgments. We do not enter the field of æsthetic judgments until we have a standard or ideal. All judgment implies a standard. Whether I say "that is beautiful," "that is right," "that is heavy," or "that is red," I affirm that the object in question comes up to the standard of my conception of what is beautiful, right, heavy, or red, as the case may be.

In the first stage or type of æsthetic judgment, the standard or ideal is undefined and undescribed. And even the critic whose business it is to express an æsthetic judgment, may not be able to define or describe the standard or ideal, and may rest content with the method of demonstration. Matthew Arnold will be admitted to be an adequate critic in questions of literary excellence; and this is what he says of those who endeavour to define the ideal in poetic work. "Critics," he writes,* "give themselves great labour to draw out what in the abstract constitutes the characters of a high quality of poetry. It is much better simply to have recourse to concrete examples; to take specimens of poetry of the high—the very highest—quality, and to say:—The characters of a high quality of poetry are what is experienced *there*. They are far better recognized by being felt in the verse of the master, than by being perused in the prose of the critic. . . . Both the substance and matter on the one hand, and the style and manner on the other, have a mark, an accent, of high beauty, worth and power. But if we are asked to define this mark and accent in the abstract, our answer must be—No, for we should thereby be darkening the question, not clearing it. The mark and accent are as given by the substance and matter of that poetry, and of all poetry which is akin to it in quality."

This then is one type of æsthetic judgment. It involves reference to a standard or ideal, but this standard is not defined in terms apprehended by the intellect; it is presented in a form which directly appeals to the emotional side of consciousness. Many of us, who would be quite at a loss to define the peculiar excellence of a good cigar or a good claret, consider ourselves none the less capable of forming a sound judgment on these important factors in the æsthetics of good living.

The second type of judgment is that of what has been termed scientific æsthetics. Here the standard or ideal is analysed, and described or defined in terms of its factors. And the judgment on any art-product is justified by the

---

* Introduction to Mr Ward's "English Poets," vol. I., pp. 27, 28.

indication of how and where, and by how much, it fails to reach the standard in question. It is, however, no part of my business here to attempt a discussion of the grounds of judgment in matters æsthetic.

Now, I think there can be no question that we may put aside this latter type of judgment as essentially human, and altogether beyond the possible powers of any animal. But with regard to the other type, there may very likely be difference of opinion as to whether animals are capable of it or not. Those who are of opinion that animals can perceive relations, will contend that they can form, though they cannot express, a judgment of this kind. Many biologists, for example, believe that birds select their mates from among numerous suitors because of their song or because of their bright plumage. Suppose a bird has two males before it, both of which are endeavouring by display of plumage, and by love-antics to win her choice. She selects the brighter, and most graceful performer. Does not this, it may be asked, imply that she has a standard of excellence, and selects that mate which she perceives as the nearer of the two to such standard? But admitting, for the purpose in hand, the correctness of the biological interpretation, that there is an exercise of choice on the part of the hen-bird, it does not necessarily follow that she perceives the relation, or compares the two competing males to an ideal standard, or even the one with the other. It is quite sufficient to suppose that A evokes a stronger emotion and a stronger appetence than B, and that she is therefore drawn to A rather than to B. There is no necessary perception of a relation, or framing of an ideal standard, of excellence. And if the facts, supposing them to be biologically well founded, can be explained on the hypothesis of sense-experience, the greater appetence prevailing, we are bound by our canon of interpretation not to assume the higher faculty of perception.

Let us note clearly what is the essential characteristic of an æsthetic judgment as such. It is essentially a matter of introspection and reflection. It is the comparison of certain emotional states aroused by *a* with the emotional states aroused by *b*, or with certain ideal emotional states which are the result of a generalization from experience, and which may be symbolized by *x*. The relation of *a* to *b*, or of *a* to *x*, is definitely perceived, and *a* is felt to approach closely to or to fall short of *b* or *x*. Primarily the æsthetic judgment is purely individual. "This is the way *a* affects me as compared with *b* or with *x*." But, apart from social life, it would be of no value to express any judgment in the matter; it would be amply sufficient to *feel* the difference. All judgments, as expressed, are of social, not of individual value. Secondarily, therefore, the judgment is social. It is an opinion expressed by me as a member of a social body, for comparison with similar opinions expressed by other members of the social body. By comparing, classifying, and generalizing these opinions, we reach the general social opinion or judgment in matters æsthetic.

Now, the result of such a comparison, classification, and generalization of opinions on questions of æsthetics differs—(1) among different individuals, (2) in different countries, and (3) at different times. This results from variation. In the matter of sense-experience we are all, with the exception of a small percentage of colour-blind, similarly affected by a red rose. In this matter, variations from a single type are not numerous or wide. But in the matter of æsthetic judgment, the variation is much more marked, and more widely spread. Hence the difficulty or impossibility of formulating a common social ideal, or standard of æsthetic judgment. Variable as it is, however, most of us do admit a social standard, even if our own individual taste does not conform to it. A man may say, "I know and acknowledge that Spenser and Milton are classics, but personally I gain no pleasure from reading their poetry." What then is this standard which we acknowledge, but may not be able personally to accept? Whose is it? Why do we acknowledge it, if it conflicts with our own individual experience?"

The standard has been reached by those who have a genuine love of literature,—to take literary judgment as an example,—who have been led to study it in all its phases, who have keen native insight and ability, and whose wide study has given them extensive experience which their ability has enabled them to utilize. If, by a practically universal consensus of opinion among those who have studied literature in this spirit, Spenser and Milton are ranked as classics of English poetry, it would be a piece of unwarrantable impertinence on the part of those of us who have less inclination towards literature, or less leisure (not to bring the question of ability into consideration at all), to set up a purely individual judgment against theirs. This, I take it, is what we mean when we acknowledge the literary excellence of work which we do not ourselves pretend to admire. And so too in other matters of æsthetic judgment.

If we call the acknowledged standard the *social standard*, we mean by this, not the average judgment of the whole social community, nor something transcending human judgment, but the average judgment of a special section of the community who have had peculiar opportunities of forming an opinion. Such social ideal or standard is based on the individual ideals or standards of those who form the special section. We take these individuals as representing the social judgment in æsthetic matters *at its best*.

I have entered into this brief discussion of the social ideal because it appears to me that the possession of ideals, æsthetic and other, and especially social ideals, is one of the distinguishing factors of the psychical life of man as compared with that of animals. Animals sense their states of consciousness as pleasure-giving or the reverse, and are largely guided in their actions by this emotional aspect of sense-experience. It is very questionable whether they perceive the relations of emotional states to each other; and if they are unable to do this, then it is obvious that they do not frame an ideal standard by reference to which the relationships shall be gauged. But if at some period in the evolution of man the perception of relations had its origin, either in the manner I have suggested or in some other way, then the more pleasureable results in any particular kind of experience would form in memory a standard in relation to which the emotional aspect of new experiences of like order would be measured. "To-day's experiences are pleasant, but not so pleasant as yesterday's." An individual standard would thus take form in the mind, and this through social inter-communication would be generalized, and become incipiently social. At length as the individual judgments of the best judges were themselves compared, and the results of comparison recorded, a widely but not universally acknowledged social standard or ideal would be reached. I say widely, but not universally, because presumably there would always be individuals who regarded the standard which they themselves framed in comparative ignorance superior to the standard acknowledged by those really competent to express an opinion. Here, therefore, as in the questions discussed in previous chapters, the turning-point in the development curve, the point of new departure, is the perception of relations.

The foregoing discussion has mainly had reference to a standard of Beauty, and to æsthetic ideals; let us now consider briefly the standard of Truth. It would be easy to quote a dozen or more instances of "deceit" on the part of animals. One which has been communicated to me by Mr Arthur Stradling must suffice. "I gave," he says, "a little Maltese terrier to a cousin of mine, who taught it, amongst other tricks, to walk round the table on its hind legs by way of earning some dessert. The dining-room and table were long, while as a rule the diners were her father and herself only. They sat at the head of the board. At the given signal the dog would start off on his bipedal journey, making the entire circuit of the table, no small task. After a while it was noticed that his speed had greatly improved; he "got home" in much less time than he had formerly occupied in completing the round. On investigation it was found that he had acquired the habit of dropping on all fours as soon as he got out of sight, running round the far end of the table, and rising again on his hind legs when he came once more into view. Furthermore, he looked back from time to time to make sure that he was not within the range of vision of his two patrons. I witnessed this little manœuvre personally, before considerations of psychological interest were ruthlessly sacrificed to a stern sense of morality, and he was broken of it."

Now the question here is, whether we are, through inability to explain this and other such cases on the hypothesis of intelligence and sense-experience, forced to assume the perception of the relation of the action, as actually performed, to the action as the master and mistress believed it to be performed. If so, the performance unquestionably rose to the dignity of a lie. We cannot here enter at length into the psychology of lying. But, whether acted or uttered, the essence of the lie is that it is used with conscious intent to deceive. The *black* that is acted or uttered has conscious reference and relation to the *white* that is not black. If the action of the Maltese terrier was performed with conscious intent to deceive, there must have passed through his mind something like that which we should thus express in words:—"I am on all fours, but they suppose I am on my hind legs." We should certainly be prepared to accept this interpretation if, *but only if*, the action cannot be interpreted on a simpler hypothesis. And here one can but regret that the action was not made the subject of an experimental investigation, and that an exact record of all the stages of the evolution of the act was not kept. I cannot, of course, say what these stages were. But *if* they were anything like the following suppositions, the action is quite explicable on the hypothesis of intelligence. The dog starts on his round on two legs, maintains this mode of progression, and earns his dessert. One day he starts on two legs, drops on all fours and comes round on all fours; no dessert. Whenever at starting he drops on all fours within sight of his mistress; no dessert. Whenever he comes in on all fours; no dessert. These are firmly associated. One day he drops on all fours, and runs round; but as he approaches the mistress the association "all fours; no dessert," leads him to jump up on his hind legs, and he earns his dessert. If once, in this or in any other such way, the dog found that by starting on two legs, running round on four, and coming in on two, he earned his dessert, this would be found to be the easiest method of doing so, and the habit would be formed and adopted. It may be said that this suggested explanation involves a somewhat complex use of intelligence. I reply that, if the action was a lie, it involved not only the complexity I have hinted at, but *in addition thereto* the thought, "while I am on all fours they suppose I am on two legs," and this is the factor which seems to me unnecessary. I believe that nearly all the anecdotes illustrative of so-called deceit in animals are explicable on the hypothesis that the animal has found, by its intelligence in sense-experience, the easiest way to do a thing or to get what it wants, and that we need not suppose that there is in the animal's mind anything of the nature of, "This is white, but master thinks it black." The "deceit" of animals (like so many of the "lies" of very little children) involves, it is true, actions which deceive or partially deceive us; but these actions are not performed with intent to deceive, or if they are the animal has got beyond the stage of sense-experience. In the language of the schools, the deceit of animals is, in my opinion, *material* and not *formal*. And I am of opinion that, as they have no perception or conception of a standard of Beauty, so they have no perception or conception of a standard of Truth. He who has no standard of truth cannot lie, though his actions may deceive us.

And what of the standard of Right? Can an animal think the *ought?* In his work on "The Human Mind,"* Professor Sully writes as follows, in a footnote: "One of the clearest examples of canine conscience I have met with," he says, "was given me by a friend, the owner of the dog, and the witness of the action. The animal, a variety of terrier, was left in the dining-room, where were the remains of a cold supper. He got on the table, and secured a piece of cold tongue, but, without eating a morsel of it, he carried it into the drawing-room, deposited it at the feet of his mistress, and then crawled out of sight, looking the picture of abject misery." I have no wish to say one word which shall detract from the moral excellence, if any one likes so to call it, of that terrier; but I may, perhaps, be allowed to analyse his case. We may suppose he was hungry, poor fellow, but the natural impulse to appease that hunger was checked in presence of that loyal feeling of subservience to the mistress which is the outcome of the dog's mode of life as the companion of man, and which has probably been developed from certain innate tendencies of the *canidæ*, which, like wolves and jackals, hunt in packs. Note, in passing, that the checking of this impulse was incomplete. Had it been complete, he would never have stolen the tongue at all, and would have remained unknown to fame. His is not the only case in which our sympathies go out towards the imperfectly moral more freely than towards those who are beyond

* Vol. ii., p. 161.

reproach. In the case of the terrier, then, the prompting of what we should call a lower impulse, the satisfaction of hunger, got the better of what we should call the higher impulse, obedience to the mistress; but only for a moment; the higher impulse prevailed, and the dog crept abjectly to his mistress. No one is likely to question—at any rate no one who knows dogs is likely to question—the existence of the higher trait in canine character, that of subservience or obedience to the master or mistress; and few are likely to question that there was in the dog's mind a painful conflict of impulses, resulting in the victory of what we call the higher. All this may be granted. But if some one says, what it should be observed Mr Sully does not say, that the terrier did what he felt to be right because he knew it to be right, that is a very different matter. That involves a thinking of the *ought;* it involves a more or less definite perception of the relation of a given act to an ideal standard. No action can be perceived to be right or wrong without reflection. The action is compared with a standard, and found either to reach or fall short of that standard. What the standard is does not matter a jot, so far as the individual moral judgment is concerned. My standard of right may be altogether wrong from my neighbour's point of view. But in expressing an individual moral judgment concerning an action, I view the act in reference to my standard, and say that it either approaches thereto or falls short thereof. Such would be the individual moral judgment. I may also compare it with the social standard, using this phrase in the sense before defined. The social standard is assuredly not the average standard of mankind. In that case it would be rather a pitiful ideal. It is the standard of the world's best and greatest and purest. Whether the comparison is with the individual ideal, or with the social ideal, it involves a perception of the relation of a given action, performed by oneself or another, to that standard. This is what I think we may, without injustice to them, deny to the brutes. The terrier's conscience, if so we may call it, on this view involved merely the emotional tone of sense-experience; it was not the moral conscience of a rational being.

The factors of the determination of conduct are several. Excluding external compulsion more or less direct, an action may be performed on prudential grounds or grounds of expediency; or it may be performed on æsthetic grounds, because the action is beautiful or seemly in the eyes of the performer or those among whom he lives; or it may be performed on moral grounds, because it is right. Often more than one factor, sometimes all the factors, co-operate in the determination of human conduct. The germs of all of these are to be found in the practical conduct of their life-activities by animals. No one will question that practical expediency is a determinant of the conduct of animals; the many recorded instances of animals checking or "punishing" others of their kind in the performance of actions which, through association of one kind or other have been rendered distasteful, appear to afford examples of the second factor; while the checking of natural propensities in subservience to the will of the beloved master would seem to illustrate the germinal phase of the third factor. But in none of these cases is there sufficient evidence to justify a belief that a standard of conduct takes form in the animal mind. The fashioning of ideals is beyond the power of the brutes. They are capable neither of regret for error on prudential grounds, nor of remorse for wrong on moral grounds. When Mr Mann Jones's dog, on whose tail he had inadvertently sat, angrily growled, and then "begged pardon for the unusual tone and temper in a way that could not be mistaken," regret is certainly implied *in the interpretation*. And Mr Jones adds: *—"Evidently he recognized his own violation

* Letter to Mr Herbert Spencer, "Justice," Appendix D, p. 277.

of an 'ought' existing in his mind (conscience)." I venture to think that the *ought* was in Mr Mann Jones's mind, not in Punch's. Similar cases must have been observed again and again by those who have gained the affection of their dumb companions. That the dog which has been forced in a moment of pain to growl at the master feels miserable thereafter, and fawns upon him with redoubled affection, is natural enough. That he recognizes his own violation of an *ought* is an explanation of the facts not in accordance with the canon of interpretation on which our method of inquiry has throughout been based. Whether we regard the end of moral endeavour as the greatest happiness of the greatest number, or as the perfecting of social integration, or as obedience to the revealed will of God, or as the realization of an ideal human self, in any case a standard of moral conduct is framed through perception and the conception that is based thereon; and actions are judged according to this standard. It is the framing of any such standard, and the perception of shortcomings therefrom, that I think we must deny to the brutes if we are to adhere to our canon of interpretation. As the animal has no ideal of Beauty, nor ideal of Truth, so too it has no ideal of Right.

It is the framing of ideals, not merely as products of conceptual thought, but also as objects of appetence and desire ever beckoning him onwards and upwards towards their realization, that is distinctive of man as man. And the worth of any individual might be gauged could we only know the nature of his ideals, the strength of his desire for their realization, and the energy and ability with which he applied himself to attain this end. In the sense-experience of animals we have the raw materials out of which all this may be elaborated. We have impressions, and the transitions between them, which need only the faculty of perception, together with the generalizing and analytic power of conceptual thought, to quicken them into knowledge. We have emotional states which, when standardized in reflection, may be sublimated to ideals. And we have a native practical energy, which needs only a new application in the Will to attain knowledge and realize the ideal. And when man became man, and began to utilize his newly acquired power, he did not leave behind him for good and all the life of sense-experience. In that life he still lives and exercises his keen intelligence. He has not abandoned the old life for a new manner of existence; he has inspired into the old life the products of higher modes of psychical activity. He is not rational in place of being intelligent; he is both intelligent and rational. He has not left behind him the emotions of his animal nature; he has idealized and purified them.

It now only remains to say a few words in conclusion. I have throughout this work accepted evolution as the basis of explanation of nature, including psychical nature. I have endeavoured to look at the facts, so far as I know them, squarely and fairly, and have not intentionally shirked the many difficulties which are incidental to the inquiry I have undertaken. I have essayed to consider mental evolution in all its aspects, and have thus been led into what some of my scientific friends will term hopelessly metaphysical speculations. But I do not think that the metaphysics of the subject can be avoided in any such inquiry. It is not a question of metaphysics or no metaphysics, but of good metaphysics or bad. In my treatment of questions of zoological psychology some will no doubt accuse me of adopting the *a priori* method; and if by the *a priori* method they mean that based on the application of general principles, I plead guilty to the charge. The question is, Have the general principles themselves been reached by the methods of scientific induction, or have they been assumed without the warranty of inductive study? I have at least done my best to make clear the grounds on which I have been led to adopt the general principles of which I have made use. Again, it may be said that throughout my discussion of zoological psychology, I have fallen into the grave philosophical error of dogmatizing from negative premisses. After making some parade of professing my inability, as human chronometer, to learn anything distinctly concerning the working of the animal clock, have I not ended by somewhat dogmatically denying to the clock certain faculties or mental powers? I have throughout been arguing, it may be said, that since I do not find sufficient evidence among animals of reason and the perception of relations, therefore such perceptual and rational powers must be absent, whereas the activities of animals can, it may be urged, be quite as well if not better explained on the assumption that they do perceive relations and exercise the faculty of reason. Such may be the line adopted by certain critics. In reply, let me say, as a last word, first, that in denying to animals the perception of relations and the faculty of reason, I do so in no dogmatic spirit, and not in support of any preconceived theory or opinion, but because the evidence now before us is not, in my opinion, sufficient to justify the hypothesis that any animals have reached that stage of mental evolution at which they are even incipiently rational; and, secondly, that I have all along based my discussion on the canon of interpretation considered in the latter part of the third chapter. If good reason can be shown for the rejection of that canon, the logical foundation of my argument will be destroyed, and the argument itself will fall to the ground.

INDEX.

ABSTRACTION, 264
Acquired characters, 347; inheritance of, 356
Adaptation, 171
Æsthetic tone, 362
Alligators, observation on, 209
Analysis, 265, 305
Anticipation, 113
Apperception, 339
Appetence and aversion, 364
Arnold, Matthew, on criticism, 365
Association, 60, 68; divergent, 69, 105; laws of, 71; in animals, 85; in the guidance of life, 88; and individual matter, 90; fibres, development of, 181; relation of, 238
Attention, 189
Auditory impressions analysed, 133
Automatism and control, 173
Automaton, 163, 175

BALDWIN, Prof. Mark, quoted, 339
Birds, effects of removing cerebral hemispheres of, 180; skill of, 161; migration of, 171, 210; song and love-antics of, 244, 363, 366; counting, 252; sense of beauty in, 362
Body and mind, 25

CANON of interpretation, 53, 242, 287, 377
Cat, observations on, 95, 297
Chameleon, sensing of distance by, 161
Chicks, observations on, 85, 94, 117, 175, 197, 209, 214, 215, 298
Child and chick, 212
Colour vision, 130
Communication in chicks, 94
Concept, 268
Conception, 263
Conceptual thought, 262
Conscience, 374
Consciousness, moment of, 11; state of, 17; curve of, 19; threshold of, 18; unity of, 21; use of term, 85; a synthetic unity, 154
Contiguity, association by, 71
Continuity of consciousness, 21, 35, 100
Contrast, association by, 79
Control, nature of, 176, 195; nerve-centres of, 178; over sensory centres, 192
Co-ordination, delicacy of, 164; nerve-centres of, 178
Correlation of sensory data, 150, 155; psychical and physiological, 152
Curiosity, 297
Curve of consciousness, 19; of molecular disturbance, 34; of reason and sense-experience, 355

DARWIN, quoted, 348

Dates as aids to time-localization, 113
Deceit in animals, 369
Depth element in vision, 148
Description, involves perception of relations, 239, 243, 295
Descriptive intercommunication, 247
Determinisim, 83, 341
Devenish, Mr, observations on alligators, 209
Differentiation in sensation, 137
Direction, sense of, 136, 171
Dog, observations on, 91, 95, 118, 159, 162, 241, 247, 254, 288, 290, 296, 300, 301, 302, 369, 372
Dualism, 1, 31
Ducklings, observations on, 85, 200, 209

EELS, migration of, 211
Effort, 196
Emotional tone, 140, 187; basis of indicative communication, 244; states, 359
Empiricism in psychology, 338
Evolution, of consciousness, 323; selective synthesis in, 333
Explanation, 275, 277, 295, 296
Extension element in vision, 150

FADING or waning of impressions, 100, 232
Fallacy, 278
Focus and margin, 15
Freedom and constraint, 340
Freewill, 189, 316
Frog, effects of removing cerebral hemispheres of, 179

GARNER, Mr R. L., observations of, 246
Gauges of mental level, 57
Geese, reported observation on, 122
Generalization, 295
Gould, Dr, on humming bird's nest, 362

HABIT, 186
Haden, Mr, observation on monkey, 303
Horse, observation on, 302
Hudson, Mr W. H., on tradition, 210
Hypnosis, 194

IDEA, 63, 269
Impressions, 63; analysis of, 125, 218; of relation, 236, subjective and objective aspect of, 309
Impulses, physiological, 28, 132
Indicative intercommunication, 244
Individuality, 220
Induction, twofold in psychology, 47
Inference, 275, 281
Infra-conscious elements, 33, 329
Inhibition, 181
Insects, psychology of sensory experience in, 158
Insight, 283
Instinct, 206, 209; deferred, 208; relation of consciousness to, 208
Intelligence, 207, 212, 215, 242, 282, 297, 371
Intercommunication, 239, 243, 293
Introspection, 20
Introspective basis of comparative psychology, 37, 44

JONES, Mr Mann, quoted, 374
Judgment, æsthetic, 364, 367

KNOWLEDGE, 119

LANCASTER, Mr, observations of, 161
Language, relation of conception to, 270
Larden, Mr, observations of, 209
Leroy, quoted, 252
Lichtenberg, on nightingale counting, 252

Life, 24, 327, 343
Living matter, characteristics of, 342
Localization in space, 123
Localization in time, 111, 120
Logic, nature of, 284
Logical object, 270; syllogism, 279
Lubbock, Sir John, observations of, 95

Margin of consciousness, 15
Materialism, 8
Medlicott, Mr H. B., quoted, 244, 248, 295
Memory, 73, 99; primary, 103; cells, 106; desultory and systematic, 108, 119
Migration, of birds, 171, 210; of eels, 211.
Mimicry and association, 97
Mind, empirical view of, 26; supplemented, 339
Moment of consciousness, 11
Monism, 1, 31, 341
Monkey, observations on, 121, 246, 253, 303
Mood, 81
Motor sensations, 142

NECESSARY truths, 278
Nerve centres co-ordinating and controlling, 176
Nerves afferent and efferent, 28, 173, 176
Nervous arc, 173
Numerical relations, 232

OBJECT, 2, 269, 310; subject and, 305, 310; of sense-experience analysed, 308
Objective study of mind, 45

PAIN, sensations of, 138
Percept, 22, 221
Perception of relations, 217; relation to conception, 263
Physical condition of vision, 128; and physiological standpoints, 230
Physiological conditions of consciousness, 24; of impressions, 127; impulses compared with sensations, 133, 144
Pigs, observations on, 245, 321
Pleasure, 140
Predication, 274
Presentative and representative, 125
Proposition, 274
Psychism, 8
Pyramidal tract, 177

QUALITIES of objects, 266, 270

REALITY, co-ordinate of subject and object, 3, 313
Reason and reasoning, 282, 287, 355
Recept, 295
Recognition, 236, 237
Reflection, 226, 234
Reflex action, 185; secondary, 186
Relations, perception of, 217; impressions of, 236; and language, 249
Relationship, association by perceived similarity of, 77
Remembering and recollecting, 111
Representative, how differentiated from presentative, 103
Resemblance, association by, 77; in animals, 96
Response due to graduated impulses, 174
Retention by ideas, 26, 106
Retentiveness, 107
Retrospection, 20, 226
Right, standard of, 372
Romanes, G. J., fetichism in animals, 92, 259; quotations from "Animal Intelligence," 122, 300, 303; observations on "Sally," 253; recepts, 295; observation on dog, 296; observations on "Cebus," 304

SATURNIA, observations on, 348
Schmankewitsch, observations of, 348
Self and not-self, 314
Sensation, a, defined, 131
Sensation, 142
Sense-experience, 157, 242, 250, 259, 265, 268, 281, 283, 355
Sense-fading, 101
Shellard, Mr E. J., quoted, 290
Similarity, association by, 76; relation of, 235
Skill, 16; of animals, 161; psychology of, 163; of chick, 198
Snakes, instinct in, 209
Social value of conception-ideal, 366
Solidity, 145
Sounds emitted by chicks, 93
Space relations, 226
Spalding, D., observations of, 202, 208
Spin, vital, 350
State of consciousness defined, 17, 125; complexity of, 156
Stimulus and response, 173
Stone, Mr, quoted, 300
Stradling, Mr Arthur, quoted, 369
Subconscious elements, 14
Subject, 2; and object, 305
Suggestion, internal and external, 61; primary and secondary, 62
Sully, Professor James, quoted, 282, 372
Syllogism, 279
Sympathy, 321
Synthesis and correlation, 144, 219, 305; selective, 333

TEMPERAMENT, 81
Therefore symbolic of logical explanation, 277; distinctive of reason, 283
Threshold of consciousness, 18
Thompson, "Passion in Animals," 252
Time-localization, 111, 120
Time relations, 230
Tradition, 170, 210
Training of animals, 93
Transitions in consciousness, 74, 221
Truth, standard of, 369

UNIFORMITY, conception of, 277; dependence of reasoning on, 277; of nature, 284
Unity of consciousness, 21

VARIATION, method of, 58; determinate and indeterminate, 345
Vision, movements in the field of, 148
Volition, 187

WALLASCHEK, Mr, on primitive music, 356
Warren, Mr R. H., quoted, 247
Wave of consciousness, 11
Weismann, Dr, on variation, 349; on musical sense, 356
Will, 188, 315
Wilson, Dr Andrew, quoted, 301
Word, relation of, to impression, percept, and concept, 271

THE WALTER SCOTT PRESS, NEWCASTLE-ON-TYNE.

THE SCOTT LIBRARY—continued.

31 VOLSUNGA SAGA. WILLIAM MORRIS. WITH INTRO-duction by H. H. Sparling.

32 SARTOR RESARTUS. BY THOMAS CARLYLE. WITH Introduction by Ernest Rhys.

33 SELECT WRITINGS OF EMERSON. WITH INTRO-duction by Percival Chubb.

34 AUTOBIOGRAPHY OF LORD HERBERT. EDITED, with an Introduction, by Will H. Dircks.

35 ENGLISH PROSE, FROM MAUNDEVILLE TO Thackeray. Chosen and Edited by Arthur Galton.

36 THE PILLARS OF SOCIETY, AND OTHER PLAYS. BY Henrik Ibsen. Edited, with an Introduction, by Havelock Ellis.

37 IRISH FAIRY AND FOLK TALES. EDITED AND Selected by W. B. Yeats.

38 ESSAYS OF DR. JOHNSON, WITH BIOGRAPHICAL Introduction and Notes by Stuart J. Reid.

39 ESSAYS OF WILLIAM HAZLITT. SELECTED AND Edited, with Introduction and Notes, by Frank Carr.

40 LANDOR'S PENTAMERON, AND OTHER IMAGINARY Conversations. Edited, with a Preface, by H. Ellis.

41 POE'S TALES AND ESSAYS. EDITED, WITH INTRO duction, by Ernest Rhys.

42 VICAR OF WAKEFIELD. BY OLIVER GOLDSMITH. Edited, with Preface, by Ernest Rhys.

43 POLITICAL ORATIONS, FROM WENTWORTH TO Macaulay. Edited, with Introduction, by William Clarke.

44 THE AUTOCRAT OF THE BREAKFAST-TABLE. BY Oliver Wendell Holmes.

45 THE POET AT THE BREAKFAST-TABLE. BY OLIVER Wendell Holmes.

46 THE PROFESSOR AT THE BREAKFAST-TABLE. BY Oliver Wendell Holmes.

47 LORD CHESTERFIELD'S LETTERS TO HIS SON. Selected, with Introduction, by Charles Sayle.

London: WALTER SCOTT, LIMITED, Paternoster Square.

THE SCOTT LIBRARY—continued.

48 STORIES FROM CARLETON. SELECTED, WITH INTRODUCTION, by W. Yeats.

49 JANE EYRE. BY CHARLOTTE BRONTË. EDITED BY Clement K. Shorter.

50 ELIZABETHAN ENGLAND. EDITED BY LOTHROP Withington, with a Preface by Dr. Furnivall.

51 THE PROSE WRITINGS OF THOMAS DAVIS. EDITED by T. W. Rolleston.

52 SPENCE'S ANECDOTES. A SELECTION. EDITED, with an Introduction and Notes, by John Underhill.

53 MORE'S UTOPIA, AND LIFE OF EDWARD V. EDITED, with an Introduction, by Maurice Adams.

54 SADI'S GULISTAN, OR FLOWER GARDEN. TRANSLATED, with an Essay, by James Ross.

55 ENGLISH FAIRY AND FOLK TALES. EDITED BY E. Sidney Hartland.

56 NORTHERN STUDIES. BY EDMUND GOSSE. WITH a Note by Ernest Rhys.

57 EARLY REVIEWS OF GREAT WRITERS. EDITED BY E. Stevenson.

58 ARISTOTLE'S ETHICS. WITH GEORGE HENRY Lewes's Essay on Aristotle prefixed.

59 LANDOR'S PERICLES AND ASPASIA. EDITED, WITH an Introduction, by Havelock Ellis.

60 ANNALS OF TACITUS. THOMAS GORDON'S TRANSLATION. Edited, with an Introduction, by Arthur Galton.

61 ESSAYS OF ELIA. BY CHARLES LAMB. EDITED, with an Introduction, by Ernest Rhys.

62 BALZAC'S SHORTER STORIES. TRANSLATED BY William Wilson and the Count Stenbock.

63 COMEDIES OF DE MUSSET. EDITED, WITH AN Introductory Note, by S. L. Gwynn.

64 CORAL REEFS. BY CHARLES DARWIN. EDITED, with an Introduction, by Dr. J. W. Williams.

London: WALTER SCOTT, LIMITED, Paternoster Square.

THE SCOTT LIBRARY—continued.

65 SHERIDAN'S PLAYS. EDITED, WITH AN INTRODUCTION, by Rudolf Dircks.

66 OUR VILLAGE. BY MISS MITFORD. EDITED, WITH an Introduction, by Ernest Rhys.

67 MASTER HUMPHREY'S CLOCK, AND OTHER STORIES. By Charles Dickens. With Introduction by Frank T. Marzials.

68 TALES FROM WONDERLAND. BY RUDOLPH Baumbach. Translated by Helen B. Dole.

69 ESSAYS AND PAPERS BY DOUGLAS JERROLD. EDITED by Walter Jerrold.

70 VINDICATION OF THE RIGHTS OF WOMAN. BY Mary Wollstonecraft. Introduction by Mrs. E. Robins Pennell.

71 "THE ATHENIAN ORACLE." A SELECTION. EDITED by John Underhill, with Prefatory Note by Walter Besant.

72 ESSAYS OF SAINT-BEUVE. TRANSLATED AND Edited, with an Introduction, by Elizabeth Lee.

73 SELECTIONS FROM PLATO. FROM THE TRANSlation of Sydenham and Taylor. Edited by T. W. Rolleston.

74 HEINE'S ITALIAN TRAVEL SKETCHES, ETC. TRANSlated by Elizabeth A. Sharp. With an Introduction from the French of Theophile Gautier.

75 SCHILLER'S MAID OF ORLEANS. TRANSLATED, with an Introduction, by Major-General Patrick Maxwell.

76 SELECTIONS FROM SYDNEY SMITH. EDITED, WITH an Introduction, by Ernest Rhys.

77 THE NEW SPIRIT. BY HAVELOCK ELLIS.

78 THE BOOK OF MARVELLOUS ADVENTURES. FROM the "Morte d'Arthur." Edited by Ernest Rhys. [This, together with No. 1, forms the complete "Morte d'Arthur."]

79 ESSAYS AND APHORISMS. BY SIR ARTHUR HELPS. With an Introduction by E. A. Helps.

80 ESSAYS OF MONTAIGNE. SELECTED, WITH A Prefatory Note, by Percival Chubb.

81 THE LUCK OF BARRY LYNDON. BY W. M. Thackeray. Edited by F. T. Marzials.

London: WALTER SCOTT, LIMITED, Paternoster Square.

THE SCOTT LIBRARY—continued.

82 SCHILLER'S WILLIAM TELL. TRANSLATED, WITH an Introduction, by Major-General Patrick Maxwell.

83 CARLYLE'S ESSAYS ON GERMAN LITERATURE. With an Introduction by Ernest Rhys.

84 PLAYS AND DRAMATIC ESSAYS OF CHARLES LAMB. Edited, with an Introduction, by Rudolf Dircks.

85 THE PROSE OF WORDSWORTH. SELECTED AND Edited, with an Introduction, by Professor William Knight.

86 ESSAYS, DIALOGUES, AND THOUGHTS OF COUNT Giacomo Leopardi. Translated, with an Introduction and Notes, by Major-General Patrick Maxwell.

87 THE INSPECTOR-GENERAL A RUSSIAN COMEDY. By Nikolai V. Gogol. Translated from the original, with an Introduction and Notes, by Arthur A. Sykes.

88 ESSAYS AND APOTHEGMS OF FRANCIS, LORD BACON: Edited, with an Introduction, by John Buchan.

89 PROSE OF MILTON: SELECTED AND EDITED, WITH an Introduction, by Richard Garnett, LL.D.

90 THE REPUBLIC OF PLATO. TRANSLATED BY Thomas Taylor, with an Introduction by Theodore Wratislaw.

91 PASSAGES FROM FROISSART. WITH AN INTRO-duction by Frank T. Marzials.

London: WALTER SCOTT, LIMITED, Paternoster Square.

GREAT WRITERS—continued.

LIFE OF ADAM SMITH. By R. B. HALDANE, M.P.

"Written with a perspicuity seldom exemplified when dealing with economic science."—*Scotsman.*

LIFE OF KEATS. By W. M. ROSSETTI.

"Valuable for the ample information which it contains."—*Cambridge Independent.*

LIFE OF SHELLEY. By WILLIAM SHARP.

"The criticisms . . . entitle this capital monograph to be ranked with the best biographies of Shelley."—*Westminster Review.*

LIFE OF SMOLLETT. By DAVID HANNAY.

"A capable record of a writer who still remains one of the great masters of the English novel."—*Saturday Review.*

LIFE OF GOLDSMITH. By AUSTIN DOBSON.

"The story of his literary and social life in London, with all its humorous and pathetic vicissitudes, is here retold, as none could tell it better."—*Daily News.*

LIFE OF SCOTT. By PROFESSOR YONGE.

"This is a most enjoyable book."—*Aberdeen Free Press.*

LIFE OF BURNS. By PROFESSOR BLACKIE.

"The editor certainly made a hit when he persuaded Blackie to write about Burns."—*Pall Mall Gazette.*

LIFE OF VICTOR HUGO. By FRANK T. MARZIALS.

"Mr. Marzials's volume presents to us, in a more handy form than any English or even French handbook gives, the summary of what is known about the life of the great poet."—*Saturday Review.*

LIFE OF EMERSON. By RICHARD GARNETT, LL.D.

"No record of Emerson's life could be more desirable."—*Saturday Review.*

LIFE OF GOETHE. By JAMES SIME.

"Mr. James Sime's competence as a biographer of Goethe is beyond question."—*Manchester Guardian.*

LIFE OF CONGREVE. By EDMUND GOSSE.

"Mr. Gosse has written an admirable biography."—*Academy.*

LIFE OF BUNYAN. By CANON VENABLES.

"A most intelligent, appreciative, and valuable memoir."—*Scotsman.*

LIFE OF CRABBE. By T. E. KEBBEL.

"No English poet since Shakespeare has observed certain aspects of nature and of human life more closely."—*Athenæum.*

LIFE OF HEINE. By WILLIAM SHARP.

"An admirable monograph . . . more fully written up to the level of recent knowledge and criticism than any other English work."—*Scotsman.*

London: WALTER SCOTT, LIMITED, Paternoster Square.

LIFE OF MILL. By W. L. COURTNEY.

"A most sympathetic and discriminating memoir."—*Glasgow Herald.*

LIFE OF SCHILLER. By HENRY W. NEVINSON.

"Presents the poet's life in a neatly rounded picture."—*Scotsman.*

LIFE OF CAPTAIN MARRYAT. By DAVID HANNAY.

"We have nothing but praise for the manner in which Mr. Hannay has done justice to him."—*Saturday Review.*

LIFE OF LESSING. By T. W. ROLLESTON.

"One of the best books of the series."—*Manchester Guardian.*

LIFE OF MILTON. By RICHARD GARNETT, LL.D.

"Has never been more charmingly or adequately told."—*Scottish Leader.*

LIFE OF BALZAC. By FREDERICK WEDMORE.

"Mr. Wedmore's monograph on the greatest of French writers of fiction, whose greatness is to be measured by comparison with his successors, is a piece of careful and critical composition, neat and nice in style."—*Daily News.*

LIFE OF GEORGE ELIOT. By OSCAR BROWNING.

"A book of the character of Mr. Browning's, to stand midway between the bulky work of Mr. Cross and the very slight sketch of Miss Blind, was much to be desired, and Mr. Browning has done his work with vivacity, and not without skill."—*Manchester Guardian.*

LIFE OF JANE AUSTEN. By GOLDWIN SMITH.

"Mr. Goldwin Smith has added another to the not inconsiderable roll of eminent men who have found their delight in Miss Austen. . . . His little book upon her, just published by Walter Scott, is certainly a fascinating book to those who already know her and love her well; and we have little doubt that it will prove also a fascinating book to those who have still to make her acquaintance."—*Spectator.*

LIFE OF BROWNING. By WILLIAM SHARP.

"This little volume is a model of excellent English, and in every respect it seems to us what a biography should be."—*Public Opinion.*

LIFE OF BYRON. By HON. RODEN NOEL.

"The Hon. Roden Noel's volume on Byron is decidedly one of the most readable in the excellent 'Great Writers' series."—*Scottish Leader.*

LIFE OF HAWTHORNE. By MONCURE CONWAY.

"It is a delightful *causerie*—pleasant, genial talk about a most interesting man. Easy and conversational as the tone is throughout, no important fact is omitted, no valueless fact is recalled; and it is entirely exempt from platitude and conventionality."—*The Speaker.*

LIFE OF SCHOPENHAUER. By PROFESSOR WALLACE.

"We can speak very highly of this little book of Mr. Wallace's. It is, perhaps, excessively lenient in dealing with the man, and it cannot be said to be at all ferociously critical in dealing with the philosophy."—*Saturday Review.*

London: WALTER SCOTT, LIMITED, Paternoster Square.

THE CANTERBURY POETS—continued.

DAYS OF THE YEAR ... With Introduction by William Sharp.

POPE ... Edited by John Hogben.

HEINE ... Edited by Mrs. Kroeker.

BEAUMONT AND FLETCHER ... Edited by John S. Fletcher.

BOWLES, LAMB, &c. ... Edited by William Tirebuck.

EARLY ENGLISH POETRY ... Edited by H. Macaulay Fitzgibbon.

SEA MUSIC ... Edited by Mrs Sharp.

HERRICK ... Edited by Ernest Rhys.

BALLADES AND RONDEAUS ... Edited by J. Gleeson White.

IRISH MINSTRELSY ... Edited by H. Halliday Sparling.

MILTON'S PARADISE LOST ... Edited by J. Bradshaw, M.A., LL.D.

JACOBITE BALLADS ... Edited by G. S. Macquoid.

AUSTRALIAN BALLADS ... Edited by D. B. W. Sladen, B.A.

MOORE ... Edited by John Dorrian.

BORDER BALLADS ... Edited by Graham R. Tomson.

SONG-TIDE ... By Philip Bourke Marston.

ODES OF HORACE ... Translations by Sir Stephen de Vere, Bt.

OSSIAN ... Edited by George Eyre-Todd.

ELFIN MUSIC ... Edited by Arthur Edward Waite.

SOUTHEY ... Edited by Sidney R. Thompson.

CHAUCER ... Edited by Frederick Noël Paton.

POEMS OF WILD LIFE ... Edited by Charles G. D. Roberts, M.A.

PARADISE REGAINED ... Edited by J. Bradshaw, M.A., LL.D.

CRABBE ... Edited by E. Lamplough.

DORA GREENWELL ... Edited by William Dorling.

FAUST ... Edited by Elizabeth Craigmyle.

AMERICAN SONNETS ... Edited by William Sharp.

LANDOR'S POEMS ... Edited by Ernest Radford.

GREEK ANTHOLOGY ... Edited by Graham R. Tomson.

HUNT AND HOOD ... Edited by J. Harwood Panting.

HUMOROUS POEMS ... Edited by Ralph H. Caine.

LYTTON'S PLAYS ... Edited by R. Farquharson Sharp.

GREAT ODES ... Edited by William Sharp.

MEREDITH'S POEMS ... Edited by M. Betham-Edwards.

PAINTER-POETS ... Edited by Kineton Parkes.

WOMEN POETS ... Edited by Mrs. Sharp.

LOVE LYRICS ... Edited by Percy Hulburd.

AMERICAN HUMOROUS VERSE ... Edited by James Barr.

MINOR SCOTCH LYRICS ... Edited by Sir George Douglas.

CAVALIER LYRISTS ... Edited by Will H. Dircks.

GERMAN BALLADS ... Edited by Elizabeth Craigmyle.

SONGS OF BERANGER ... Translated by William Toynbee.

HON. RODEN NOEL'S POEMS. With an Introduction by R. Buchanan.

SONGS OF FREEDOM. Selected, with an Introduction, by H. S. Salt.

CANADIAN POEMS AND LAYS ... Edited by W. D. Lighthall, M.A.

CONTEMPORARY SCOTTISH VERSE. Edited by Sir Geo. Douglas.

POEMS OF NATURE ... Edited by E. W. Rinder.

CRADLE SONGS ... Edited by Grace Rhys.

Printed in Dunstable, United Kingdom

91219927R00238